Molecular and Physiological Insights into Plant Stress Tolerance and Applications in Agriculture

Edited by

Jen-Tsung Chen
Department of Life Sciences
National University of Kaohsiung
Taiwan

Molecular and Physiological Insights into Plant Stress Tolerance and Applications in Agriculture

Editor: Jen-Tsung Chen

ISBN (Online): 978-981-5136-56-2

ISBN (Print): 978-981-5136-57-9

ISBN (Paperback): 978-981-5136-58-6

need for a court order if at any point you breach any terms of this License Agreement. In no event will any delay or failure by Bentham Science Publishers in enforcing your compliance with this License Agreement constitute a waiver of any of its rights.

3. You acknowledge that you have read this License Agreement, and agree to be bound by its terms and conditions. To the extent that any other terms and conditions presented on any website of Bentham Science Publishers conflict with, or are inconsistent with, the terms and conditions set out in this License Agreement, you acknowledge that the terms and conditions set out in this License Agreement shall prevail.

Bentham Science Publishers Pte. Ltd.
80 Robinson Road #02-00
Singapore 068898
Singapore
Email: subscriptions@benthamscience.net

BENTHAM SCIENCE

CONTENTS

FOREWORD

Stress tolerance is a continuing issue for researchers and professionals seeking to increase crop productivity. In the research field of plant science, stress physiology is an intensive topic for researchers, and tons of publications are reported per year to get increasing knowledge about stress tolerance when facing global climate change. In the meantime, the emerging knowledge of plant stress physiology should be applied to the practice of agriculture for sustainable agriculture as well as food security globally. Importantly, there is a high demand for the integration of current knowledge of plant stress physiology. Moreover, a systematic summary of methods in plant stress management also needs to be refined.

This book, "Molecular and Physiological Insights into Plant Stress Tolerance and Applications in Agriculture," collects the most recent original research and literature reviews for unraveling the physiology of plant stress tolerance. Divided into 21 chapters, it provides in-depth coverage of the recent advances by exploring the unique features of stress tolerance mechanisms, which are essential for better understanding and improving plant response, growth, and development under stress conditions, in particular by exploring knowledge that focuses on the application of plant growth regulators, advanced biotechnologies, high-throughput technologies, multi-omics, bioinformatics, systems biology, and artificial intelligence as well as beneficial microorganisms on the alleviation of plant stress.

The mechanisms covered in this book include the perception of stress, signal transduction, and the production of chemicals and proteins associated with the stress response. The book also offers critical knowledge of the gene networks involved in stress tolerance and how they are used in plant stress tolerance development. Modern genetic studies and useful breeding methods are also covered. It also presents the current challenges and further perspectives. Therefore, this book might largely benefit breeding programs as well as sustainable agricultural production in the future.

The editor, Prof. Jen-Tsung Chen has done an excellent job of bringing together specialists from diverse fields to present the most comprehensive view of current research findings and their implications for plant stress tolerance physiology. Therefore, this book, "Molecular and Physiological Insights into Plant Stress Tolerance and Applications in Agriculture," is an essential resource for academics and professionals working in agronomy, plant science, and horticulture. It is an essential resource for both novices and specialists. It can also be utilized as a resource for courses at the university level for students and Ph.D. students interested in the physiology of plant stress tolerance. I recommend it without reservation!

Christophe Hano
University of Orleans
France

PREFACE

This book is an edited collection of a series of chapters presented by experts in the cross-disciplinary research fields of plant physiology, plant stress, agronomy, agriculture, horticulture, microbiology, molecular biology, multi-omics, plant pathology, and environmental science. To ensure its broad sources of knowledge and transparency, over one thousand invitations were sent by email to experts all over the world together with the announcement of a call for papers in the ResearchGate which attracted over seven hundred "reads" and twenty-five "followers" for about half a year of posting, more than thirty groups express their interest in contributing chapters to this book after two months of communication. The content of the chapters was finished after four to five months of preparation and eventually, twenty-one of them were accepted for publication. So, I would like to thank all the contributors to this book for their time and effort in organizing valuable chapters.

This book consists of five major subtopics, including 1) the fundamental theory of molecular biology and plant stress physiology; 2) microbial dynamics within the rhizosphere of plants and the use of plant growth-promoting bacteria in the improvement of growth and productivity of crops; 3) morphological and physiological responses of plants and the underlying molecular mechanisms under abiotic stress particularly salt, heat and drought, and importantly, several chapters pay their attention to systematic describe molecular aspects of salt tolerance in plants; 4) the current applications of biotechnology and the benefits to sustainable agriculture; 5) the techniques for improving tolerance to biotic stress such as the use of endophytic bacteria.

The contents of this book could enrich our understanding of molecular and physiological mechanisms in plant abiotic and biotic stresses and their interactions. It collects the recent advances through refining the knowledge systematically from a large amount of literature of more than two thousand and five hundred publications and organizes thirty-seven figures and thirty-five comprehensive tables to attract readers. This book could be a critical reference and suitable textbook for researchers, students, teachers, growers, and experts in a broad range of fields such as plant science, agronomy, and agriculture. It contributes to the significant progress in expanding our knowledge in the field of plant stress physiology and the contents benefit in obtaining stress-tolerant crops; enhancing their quality and productivity; and eventually, supporting the sustainability of agriculture and the global food supply.

This book has been divided into two parts. In part I, the emerging knowledge on plant abiotic/environmental stress tolerance was included, which covers stressors including temperature, salt, and drought and their combinations. When faced with global climate change, these abiotic stresses can be worse chiefly due to a rise in temperature. Firstly, the fundamental theory, methods, and applications of stress mitigation technology have been refined in this book. It provides an in-depth discussion on salt stress response and its mitigation strategies, particularly, several chapters focus on recent achievements in significant crops such as sugar beet and rice. In addition, temperature and drought stresses were systematically reviewed and the authors give perspective for the future directions. Part I of this book aims to support the goals of SDGs (Sustainable Development Goals), particularly achieving zero hunger through enhancing agricultural production.

Jen-Tsung Chen
Department of Life Sciences
National University of Kaohsiung
Taiwan

List of Contributors

Abhishek Kanojia	Department of Botany, University of Delhi, New Delhi-110007, India
Anamika Barman	Division of Agronomy, ICAR- India Agricultural Research Institute, New Delhi-110012, India
Anurag Bera	Department of Agronomy, Institute of Agricultural Sciences, Banaras Hindu University, Varanasi, Uttar Pradesh, 221005, India
Ashutosh Kumar Mall	ICAR-Indian Institute of Sugarcane Research Lucknow-226 002, (U.P.), India
Asma Shakeel	Faculty of Agriculture, SKUAST, Kashmir, 193201, India
Ayushi Jaiswal	Department of Botany, University of Delhi, New Delhi-110007, India
B. Umakanth	Plant Breeding Department, ICAR-Indian Institute of Rice Research, Hyderabad 500030, India
Bharti Choudhary	IILM University, Greater Noida, Uttar Pradesh 201306, India
Battana Swapna	Department of Botany, Vikrama Simhapuri University College, Kavali-524201, SPSR Nellore District, Andhra Pradesh, India
Chanchal Kumari	Laboratory of Plant Functional Genomics, Regional Centre For Biotechnology, Faridabad, India
G. Padmavathi	Plant Breeding Department, ICAR-Indian Institute of Rice Research, Hyderabad 500030, India
G.S.V. Prasad	Plant Breeding Department, ICAR-Indian Institute of Rice Research, Hyderabad 500030, India
Inayat M. Khan	Faculty of Agriculture, SKUAST, Kashmir, 193201, India
Jitendra Singh Bohra	Department of Agronomy, Institute of Agricultural Sciences, Banaras Hindu University, Varanasi, Uttar Pradesh, 221005, India
K. Muralidharan	Pathology Department, ICAR-Indian Institute of Rice Research, Hyderabad 500030, India
M.N. Arun	Agronomy Department, ICAR-Indian Institute of Rice Research, Hyderabad 500030, India
Mehnaz Shakeel	Faculty of Horticulture, SKUAST, Kashmir 190025, India
Mandala Ramakrishna	Department of Botany, Vikrama Simhapuri University College, Kavali-524201, SPSR Nellore District, Andhra Pradesh, India
Neha Sharma	IILM University, Greater Noida, Uttar Pradesh 201306, India
Nimisha Sharma	Indian Agricultural Research Institute, New Delhi, India
Nesreen H. Abou-Baker	Soils and Water Use Department, National Research Centre, Dokki, Giza, Egypt
Priyanka Saha	Division of Agronomy, ICAR- India Agricultural Research Institute, New Delhi-110012, India
R. K. Singh	International Centre for Biosaline Agriculture Dubai, United Arab Emirates

Ramu S. Vemanna	Laboratory of Plant Functional Genomics, Regional Centre For Biotechnology, Faridabad, India
Ravi Rajwanshi	Discipline of Life Sciences, School of Sciences, Indira Gandhi National Open University, New Delhi, India
Shobhna Yadav	Laboratory of Plant Functional Genomics, Regional Centre For Biotechnology, Faridabad, India
Syed Andleeba Jan	Faculty of Agriculture, SKUAST, Kashmir, 193201, India
Shakeel A Mir	Faculty of Agriculture, SKUAST, Kashmir, 193201, India
Sekhar Tiwari	School of Sciences, P P Savani University, Surat, Gujarat, India
Srinivasan Kameswaran	Department of Botany, Vikrama Simhapuri University College, Kavali-524201, SPSR Nellore District, Andhra Pradesh, India
Thummala Chandrasekhar	Department of Environmental Science, Yogi Vemana University, Kadapa-516005, Andhra Pradesh, India
Varucha Misra	ICAR-Indian Institute of Sugarcane Research Lucknow-226 002, (U.P.), India
Yashwanti Mudgil	Department of Botany, University of Delhi, New Delhi-110007, India
Z. Mehdi	Faculty of Agriculture, SKUAST, Kashmir, 193201, India

CHAPTER 1

Influence of Abiotic Stress on Molecular Responses of Flowering in Rice

Chanchal Kumari[1], **Shobhna Yadav**[1] and **Ramu S. Vemanna**[1,*]

[1] *Laboratory of Plant Functional Genomics, Regional Centre For Biotechnology, Faridabad, India*

Abstract: Rice is a short-day plant, and its heading date (Hd)/flowering time is one of the important agronomic traits for realizing the maximum yield with high nutrition. Theoretically, flowering initiates with the transition from the vegetative stage to shoot apical meristems (SAMs), and it is regulated by endogenous and environmental signals. Under favorable environmental conditions, flowering is triggered with the synthesis of mobile signal florigen in leaves and then translocated to the shoot for activation of cell differentiation-associated genes. In rice, the genetic pathway of flowering comprises *OsGI–Hd1–Hd3a,* which is an ortholog of the *Arabidopsis GI–CO–FT* pathway, and the *Ehd1-Hd3a* pathway. Climate change could affect photoperiod and temperature, which in turn influences heading date and crop yield. In low temperatures and long-day conditions, the expression of the *HD3a* gene analogous to *FT* in *Arabidopsis* deceased, which delays flowering. Similarly, during drought, expression of the *Ehd1* gene is suppressed, resulting in a late-flowering phenotype in rice. Drought affects pollen fertility and reduction in grain yield by reducing male fertility, which affects male meiosis during reproduction, microspore development, and anther dehiscence. In this research field, substantial progress has been made to manipulate flowering-related genes to combat abiotic stresses. Here, we summarize the roles of a few genes in improving the flowering traits of rice.

Keywords: Abiotic stress, Flowering, Florigen, *Hd1*, *Hd3a*, Long-day, Short-day.

1. INTRODUCTION

Rice is an important staple food crop, consumed by more than 50% of the population in world. Rice is grown in puddled conditions in general, however, recently, aerobic rice cultivation has been gaining importance in saving water. Drought is one of the most severe forms of abiotic stress affecting plants, due to global climate change. Due to the high requirement for water, the rice crop is severely affected during drought conditions. Drought negatively affects plant

[*] **Corresponding author Ramu S. Vemanna:** Laboratory of Plant Functional Genomics, Regional Centre For Biotechnology, Faridabad, India; E-mail: ramu.vemanna@rcb.res.in

Jen-Tsung Chen (Ed.)

growth and grain quality due to hindrances in the physiological and metabolic processes, including respiration, photosynthesis, stomatal opening-closing, *etc*. The Rice plant takes around 3-6 months to grow and consists of vegetative, reproductive, and ripening stages. The reproductive stages in rice begin with booting, which leads to a bulging of the leaf, the stem that conceals the developing panicle, followed by the emergence of the stem that continues to grow. Flowering or heading in rice begins when the panicle is fully grown [1, 2]. Drought affects rice growth at almost every stage, most severe at flowering, followed by booting and grain-filling stages. Drought affects spikelet fertility at the panicle initiation stage, resulting in low grain yield [3, 4]. Another crucial environmental factor affecting plant growth and reproduction is extreme temperatures, leading to a significant decrease in crop yield. An increase in temperature induces floret sterility in rice due to another indehiscence by interfering with pollen grains swelling [1, 2]. Similarly, cold stress in the range of 15–19°C, significantly affects the reproductive stage and reduces rice yield. The reduced temperature at the heading stage in rice causes improper development of microspore, which produces infertile pollen grains, resulting in high spikelet sterility and low nutritional quality grains [3].

Being sessile, plants have evolved many stress adaptation and escape mechanisms. *Arabidopsis* has evolved a drought escape (DE) mechanism by shortening its life cycle to avoid drought [5]. However, when rice is exposed to water-deficit conditions, it can trigger early flowering and reduced tiller numbers or delay in flowering, depending upon drought severity [5]. Reduction in tissue water status during drought stress triggers various signaling molecules such as calcium influx from the extracellular matrix to cytosol, inositol-1, 4, 5-triphosphate (IP3), abscisic acid (ABA), cyclic adenosine diphosphate ribose (cADPR), nitric oxide, *etc*. Secretion of these signaling molecules alters diverse signaling pathways, which play a critical role in determining the flowering and productivity. The environmental cues and these signaling molecules are perceived by a diverse family of receptors, which activate or suppress a cascade of signaling events. Upon perception, receptor-ligand interactions trigger the activation of kinases which affect the expression of diverse transcription factors such as HD-zip/bZIP, AP2/ERF, NAC, MYB, WRKY, and other genes [6 - 10]. The cell fate-determining factors play an important role in flower initiation and transition mechanisms. The mitogen-activated protein kinases (MAPKs), proline (pro), late embryogenesis abundant (LEA), glycine betaine (GB), soluble sugar (SS), proteins and aquaporin (AQP) are upregulated during drought, which favors plant survival in harsh conditions [11]. To develop a climate-resilient rice crop, which can withstand adverse abiotic stresses, it is important to understand the plant responses to environmental factors. Efforts have been made to understand the molecular mechanisms involved in flowering, and attempts have been made to design climate-resilient crops.

1.1. Receptor for Light and Temperature

To adapt to climatic conditions, plants have evolved different photoreceptors such as cryptochrome (*CRY*), phytochrome (*Phy*), and phototropin (*PHOT*) for sensing light which regulates photoperiod (day length). Cryptochromes are flavin-containing photoreceptors that interact with a DNA repair enzyme photolyase, which gets oxidized and activated by blue and UV-A light, leading to the formation of a semiquinone intermediate that can efficiently absorb green (500–600 nm) light. In rice, two cryptochrome genes are present: OsCRY1a and OsCRY2 (formerly known as OsCRY1b), which promote rice flowering time [12, 13]. Mutations in these genes inhibit coleoptile and leaf elongation upon blue light irradiation [12]. *OsCRY1a* and *OsCRY1b* inhibit rice seedling development by physically interacting with Constitutively Photomorphogenic 1 (COP1) protein that inhibits coleoptile development and leaf elongation [12, 14]. *COP1* encodes RING-finger E3 ubiquitin ligase and interacts with the Suppressor of *phyA-105* (SPA) to degrade CONSTANS (CO), a central regulator of flowering. *CO* expression is indirectly regulated by the degradation of GIGANTEA (GI) [15]. Phototropins have two flavin binding domains at the N-terminal and a serine/threonine kinase domain at the C-terminal end. *PHOT1* and *PHOT2* regulate phototropism and stomatal movement in plants. Phytochromes are the best-known blue-green pigment receptors, ranging from PhyA to PhyC, which sense different wavelengths of light from red to far-red. These are cytoplasmic, dimeric, serine/threonine kinases activated in response to red light by absorbing 560 nm wavelength converting inactive Pr (Red) form to active Pfr (far-red) form and Pfr to Pr by absorbing far-red light of ~730 nm wavelength. The inactive form of Pr is present in cytosol as it converts to the active Pfr form translocated to the nucleus, where it interacts with multiple partners to influence the expression of many target genes involved in photomorphogenesis [12]. Phytochromes play a critical role in the process of anthesis, influencing the heading date and overall growth of rice plants. Double mutants of *phyA* and *phyB* show defects in pollen development and anthesis [14]. Phytochromes also play an important role in flowering by regulating the expression of *Hd3a* in rice *via Hd1*. Hd1, in turn, is regulated at the post-translational level by PhyB [16]. PhyB also acts as a thermosensor, and its direct interaction with the promoter region of key target genes in a temperature-dependent fashion has been reported. Phytochromes are involved in the convergence of light and temperature in regulating photomorphogenesis [16].

Phytochromes interact with the basic helix-loop-helix (bHLH) group of transcription factors, *i.e.*, Phytochrome Interacting Factors (PIFs), for maintaining the skotomorphogenic state (development of seedlings in the dark) of plants in dark condition. Phytochromes are activated upon exposure to light and interact with PIFs, leading to phosphorylation, ubiquitination, and proteasome-mediated degradation, which results in photomorphogenic development [14]. PIFs interact with the promoter of circadian genes to regulate day-length determinants [17]. The phytochrome-interacting factor-like genes (*OsPIL11* to *OsPIL16*) have been identified, and the *OsPIL1/OsPIL13* provides drought stress tolerance by influencing cell wall-related genes. Expression of *OsPIL1/ OsPIL3* leads to reduced internode elongation and enables them to grow under shade from neighboring plants [18]. The role of *PIL15* has been identified in grain development, which negatively affects cytokinin transport which is essential for cell division during the grain-filling stage. *OsPIL15* binds to the N1-box (CACGCG) motifs, present in the *OsPUP7* promoter of the purine permease gene, which can transport caffeine, a CK derivative. Overexpression of *OsPIL15* in rice endosperm leads to small grain size, and CRISPR-Cas9 mutant lines showed an increase in grain size due to reduced expression of *OsPUP7*. The reduced levels of *PUP7* interfere with cytokinin transport from other parts of the plants to the grain [19].

1.2. Reproduction and Maintenance of Shoot Apical Meristem

Rice spikelet consists of two pairs of lemmas and rudimentary glumes, leaf-like structures, and a single fertile flower that comprises a palea, a lemma, one pistil, two lodicules, and six stamens Fig (**1**). Seed production and inflorescence architecture in rice are primarily determined by spikelet morphogenesis [20]. Plant development initiates from the fertilization of egg cells with sperm nuclei to form a zygote, and majorly includes embryogenesis, vegetative, and reproduction phase. During the embryogenesis stage, the zygote divides without any morphogenetic event for plant development. The mature rice embryo is composed of scutellum, coleoptile, epiblast, radicle and shoot apical meristem (SAM). After seed germination, SAM develops above-ground parts of the plant by providing a perpetual supply of cells [21]. SAM is regulated in rice by a feedback loop of *WUSCHEL-RELATED HOMEOBOX4* (*WOX4*), *FLORAL ORGAN NUMBER-1* (*FON1*), and *FLORAL ORGAN NUMBER-2* (*FON2*). *FON1* and *FON2* are closely related to the *CLAVATA1* (*CLV1*) and *CLAVATA3* (*CLV3*) of Arabidopsis involved in the transition between the shoot and inflorescence meristem [14]. *FON-2*, like *CLE 1,2* (*CLAVATA3*/Endosperm surrounding region-related protein-*FCP1, FCP2*), is analogous to *WUSCHEL* and *CLAVATA* in *Arabidopsis*. The co-silencing of *FCP1* and *FCP2* in rice increases the meristem activity, suggesting its role as a negative regulator of meristem maintenance. In contrast, silencing of *WOX4* leads to a reduction in SAM size, showing small abnormal shoots with yellow-colored leaves. The function of *WOX4* is reported to play a role in cytokinin accumulation in SAM [22]. Mutations in *Shoot Organization* (*SHO1*, *SHO2*), *small RNA SHOOTLESS 2* (*SHL2*), and a homolog of *Arabidopsis DICER* *like-4*, *AGO7*, and RNA dependent RNA polymerase 6 (RdRP) caused severe SAM inhibition [13, 23]. MicroRNAs also showed an important role in phase transition in SAM; *Osa-miR156* binds to *SQUAMOSA PROMOTER BINDING PROTEIN LIKE* (*SPL*), which in turn affects the expression of MADS-box genes and leads to the transition of the vegetative stage into reproductive stage [24]. In delayed heading date (*dh*), mutant upregulation of *miRNA-OsMIR171c*, target GRAS (*GAI-RGA-SCR*), and other plant-specific transcription factors (*OsHAM1*, *OsHAM2*, *OsHAM3*, and *OsHAM4*) resulted in prolonged vegetative stage along with other phenotypic defects. These *dh* mutants also showed up-regulation of *Osa-miR156*, which suppressed expression of the flowering integrators *Hd3a* and *RFT1* [24].

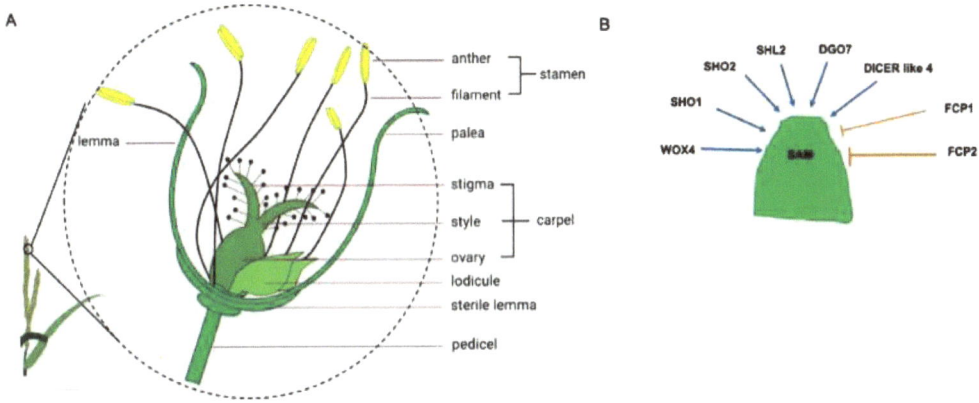

Fig. (1). Schematic representation of floral parts of rice and the genes involved in shoot apical meristem maintenance. **A**. Different parts of a flower in a panicle. **B**. Genes involved in the regulation of Shoot Apical Meristem (SAM). orange arrow -negative regulator and blue arrow - positive regulator of SAM.

1.3. Molecular Mechanisms of Flowering

Flowering in plants is controlled by environmental and endogenous signals. The transition of the reproductive stage from the vegetative stage is one of the critical developmental steps in plants. Based on the day length duration perceived, plants are categorized into three major classes: long-day (LD), short-day (SD), and day-neutral. Rice is a short-day plant, and flowering time is controlled by a complex network of multiple genes [25]. The effect of vernalization has no influence on flowering in rice [26]. However, photoperiod is the most important environmental factor for flowering signaling. The Rice Indeterminate 1 (*RID1*), encodes a transcription factor Cys-2/His-2-type zinc finger and acts as a master regulator in the transition from SAM to the flowering stage. The *rid-1* mutant showed a never flowering phenotype, by reducing the expression of florigen, *Hd3a*, and *Ehd1* genes under SD and LD conditions [27]. The expression of *Hd1* and *Ghd7* was reduced in *rid-1* plants, showing partial regulation of both genes by *RID1*.

However, expression of flowering repressor *RICE CENTRORADIALIS 1(RCN 1)*, an important gene in the transition from SAM to the flowering stage, was not influenced in *rid-1* plants, showing both of them functioning independently [27].

Flowering in rice is regulated by two pathways, one involving *Hd1-Hd3a* and another evolutionarily unique pathway is mediated by *Ehd1*, with no orthologs in *Arabidopsis*, under both SD and LD by modulating expression of the rice florigen genes *Hd3a* and *Rice Flowering Locus-T1 (RFT1)*. The day length is perceived by leaves which induces the expression of florigens, *Hd3a*, and *RFT1*. Further, these florigen proteins move from leaves to SAM *via* the plant vascular system and initiate the onset of flowering Fig (**2**) [28, 29]. In SAM, Hd3a interacts with floral differentiation 1 (FD1), a bZIP transcription factor *via* intracellular receptor 14---3, and acts as a bridge in the formation of the florigen activation complex (FAC). Formation of FAC complexes comprising Hd3a-OsFD1-14-3-3 in SAM activates transcription of the target genes *APETALA (AP1)/FRUITFULL (FUL)-like* genes *OsMADS14*, *OsMADS15* and *OsMADS18,* which are involved in flowering initiation [30]. Suppression of these genes with RNAi showed a delay in floral transition. These *MADS* genes function in SAM, along with *PANICLE PHYTOMER2 (PAP2)* from the *SEPALLATA* subfamily. The quadrupole mutant displayed continuous development of the leaf, instead of inflorescence meristem [31].

Expression of *Heading date (Hd3a)* is responsive in short-day conditions, and it decreases strikingly under day length longer than 13.5 h. During short-day conditions (SDs), *Hd1* acts as a key regulator in promoting flowering by regulating the expression of *Hd3a* and inhibiting under long-day conditions (LDs). The expression of *Hd1* is regulated by *OsGIGANTEA (GI),* a key component for perceiving the circadian clock and light signals. GI regulates the expression of the *OsMADS51* gene involved in lemma differentiation and regulates the expression of *EHD1, Hd3a,* and *OsMADS14* [32]. Unlike *Arabidopsis*, a mutation in *OsGI,* does not affect the expression of genes involved in primary metabolite and maintains higher photo-assimilates. However, phenylpropanoid metabolite pathway genes were altered consistently [33]. HD1 is activated upon phosphorylation by Heading Date Repressor 1 (*HDR1*), a flowering repressor [34]. This leads to the upregulation of *HD1* and downregulation of *Hd3a* and *RFT1* under LD, suggesting a role of *Hd1* in regulating flowering time *via* the photoperiodic pathway. *Ehd1* mediates another evolutionary unique pathway in rice flowering as its orthologue is not present in *Arabidopsis*. *Ehd1* promotes flowering in rice under both SD and LD by promoting the rice florigen genes *Hd3a* and *RFT1* [29, 34]. Flowering under long-day conditions is regulated by Days to Heading on chromosome 2 (*DTH2*), which influences the expression of *Hd3a* and *RFT1* more independently than *Ehd1* [35].

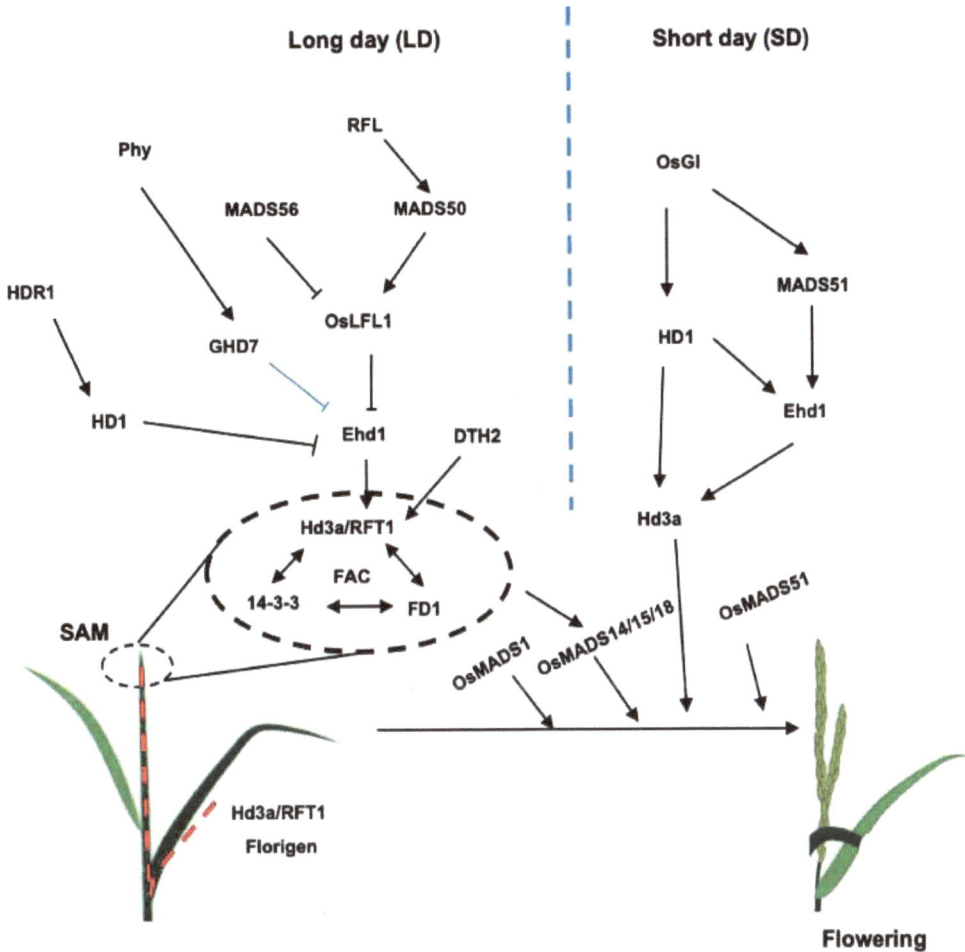

Fig. (2). Pathways showing genes involved in flowering under long-day (LD) and short-day (SD) conditions. The transition of shoot apical meristem to inflorescence meristem is mediated by the *Hd3a/RFT* gene *via* mainly two different pathways, *Hd1-Hd3a* and others involving *Ehd1* as intermediate. Under LD conditions, *HD1* suppresses the *Ehd1* expression, while in SD conditions, it upregulates *Ehd1*. The flower activation complex (FAC) is formed by the interaction between Hd3a - 14-3-3 - FD1 upregulating the *OsMADS14/15/18*, converting shoot apical meristem to inflorescence meristem.

RICE FLORICULA/LEAFY (RFL), *Arabidopsis* orthologue of Leafy (LFY), plays an important role in flowering time and whole plant architecture maintenance. The *RFL* overexpression lines trigger precocious flowering by increasing the expression of flowering activator *OsMADS50, an* orthologue of the *SOC* gene in *Arabidopsis* [36]. *OsSOC1* functions parallel to *Hd1* and *OsGI*, which regulate the expression of *FT*-like *Hd3a via the LFY* gene. In Arabidopsis, activation of *HD1*, *SOC1, LFY,* and *FT* occurs consecutively. However, in rice, *RFL* acts upstream of *OsSOC1/ OsMADS50* and *RFT1* (rice florigen) to promote flowering [36]. While *OsMADS56* repress the flowering under long-day condition *via* the *OsLFL1-EHD1* pathway [37]. *RFL* is a regulator of several unique genes encoding transcription factors, ethylene signaling factors (*EIL3*), auxin efflux facilitator (*PIN3*-like), and hormone biogenesis/catabolism genes [36]. Under drought stress, *RFL* is induced and responsible for early flowering [17]. A bZIP transcription factor ABA-responsive element binding factor1 (*OsABF1*) is also induced, which indirectly reduces the transcription level of *Ehd1* by activating the *OsWRKY104* leading to the accumulation of its upstream genes, *Hd3a* and *RFT1*. The *Ghd7* gene is also reduced due to drought stress, irrespective of the photoperiodism [18, 19]. In rice, Jasmonic acid regulates rice spikelet development by *Extra Glume 1* (*EG1*) encoding phospholipase, which catalyzes the first step of JA biosynthesis with the release of linolenic acid (C18:3) from chloroplast membrane lipids. The *eg1* and *eg2* mutants had defective floral meristem determinacy and showed altered spikelet morphology and floral organ identity. JA biosynthesis is mediated by *EG1*, whereas *EG2/OsJAZ1* act as JA signalling repressor. *OsJAZ1* interacts with *OsCOI1b* (JA receptor) to trigger self-degradation and additionally, it interacts with *OsMYC2* to suppress its role in *OsMADS1* activation, an E-class gene playing important role during rice spikelet development. The role of *OsMADS1* is reported in the maintenance of determinate floret development by ovule and carpel differentiation regulating determinate floral development [20].

1.4. Adaptation of Rice to Different Climatic Conditions

Thousands of years ago, wild relatives of rice started growing in tropical conditions at 28 ˚C North latitude. But, with the process of domestication, they started spanning temperate areas with the difference in photoperiod and cold temperature. The expansion of rice in the northern hemisphere was possible through domestication and breeding techniques, due to reduced flowering time, and less sensitive photoperiod trait, which is essential for harvesting produce before cold commence [20]. *Hd1* expression leads to flowering in rice under short-day conditions. Another unique gene *Ehd1* induces flowering in rice by activating short-day florigen *Hd3A* or long-day florigen *RFT1* under short/ long-day conditions, respectively. *Ehd1* promotes flowering in the rice harboring the non-functional allele of *Hd1*, by activating *Hd3a,* which functions redundantly. However, in long-day conditions, *a* functional allele of *Hd1* could not complete the role of *Ehd1,* as the non-functional *Ehd1* allele containing rice did not flower even after 180 days, showing the antagonistic role of *Hd1* and *Ehd1* under LD. These two independent flowering pathways work together for the adaptation of rice to different climatic conditions [38, 39]. The rice cultivars growing in the temperate zone, with long photoperiod in most of the year, have a functional allele of *Hd1* that favors early flowering before cold commences. However, rice growing in tropical/sub-tropical conditions have non-functional *Hd1* that favors extended heading date [40]. In line with this functional *RFT1*, long-day florigen is mostly present in Indica cultivars grown at higher latitudes in the temperate zone. The non-functional *RFT1* was found in some Indica rice varieties grown at low latitudes with a tropical climate. The functional *RFT1* exists in long-day florigen in Japonica and Indica cultivars in temperate/high latitudes, while non-functional *RFT1* is detected in some Indica rice varieties in lower latitudes/tropical climates [29].

Expression of *Gdh7*, a quantitative trait locus (QTL) encoding for CCT domain (CO, CO-LIKE, and TIMING OF CAB1) controlling multiple traits, including flowering time, plant length, or the number of grains per panicle varies for adaptation to the different climatic zone. In long-day conditions, increased expression of *Ghd7* delays flowering time and increased height and panicle size in rice. For adapting to get cultivated in temperate regions, natural mutants with reduced function have evolved with time, enabling rice cultivars to grow in cooler temperatures. Thus, *GhD7* plays a crucial role in rice production and adaptation globally [41]. Transcriptome analysis of cold-sensitive Japonica variety Yongyou 538, showed some differentially expressed genes with upregulation of key flowering genes such as *OsGI, OsFKF1, Hd1, FT*-1, and *OsELF3*-2, in comparison to insensitive variety Niggeng 4. This study shows changes in the expression of flowering genes for regulating the adaption of rice crops to different environmental conditions [42].

1.5. Development Made at Molecular Level to Combat Abiotic Stress in Plants

Drought stress under both long and short-day photoperiods affects flowering in rice. Rice uses drought escape (DE) response to combat drought stress at the vegetative stage and extends the heading date [5]. Although drought affects the development of both female and male reproductive organs, severity is associated with male organ development. Editing of uORFs in 5' leader sequence of *Hd2*, a flowering repressor extended heading date with 4-12 days, with reduced expression of *RFT1, Ehd1,* and *Hd3a*. Delay in heading date has the potential for adaptation of rice cultivars from tropical to temperate regions and confer drought escape mechanism [43]. Tolerance to drought stress has been achieved by targeting *Drought-Induced LTP (OsDIL1)* gene, encoding for lipid transfer protein. Expression of this gene is reported to be higher in the meiotic and post-meiotic flowering stage in comparison to other tissues, suggesting its role in flower development. The *OsDIL*-overexpression lines showed reduced expression of *OsCP1, OsC4,* and *CYP704B2,* another developmental gene in response to drought, showing a role in pollen fertility. Overexpression plants of *OsDIL* also showed higher levels of ABA synthesis gene *ZEP1* (*Zeaxanthin epoxidase*) and other drought-related genes, including *basic leucine zipper* (*bZIP46*), *RESPONSIVE TO DEHYDRATION22 (RD22), SOD1 (Superoxide dismutase)*, and *peroxidase* (*POD*) stress marker genes [44].

In plants, R2R3-MYBs have been extensively characterized for providing tolerance to various abiotic stress. The MYB Important for *Drought response 1* (*MID1*) encoding putative R-R-type MYB-like transcription factor improves rice yield under drought conditions. *MID1*-overexpressing plants showed higher seed settings due to fertile pollen development. *MID1* acts as a transcriptional regulator of floret development in rice, promoting the development of male reproductive organs under drought conditions. *MID1* combats abiotic stress by binding to *β-ketoacyl reductase* gene, required for the first step of the fatty acid synthesis, and rice *Male Sterility2* gene (*MS2*) important for anther and microspore development [21]. The *MID-1* binds to the CYP707A5 promoter, regulating ABA catabolic pathway, leading to a decrease in ABA levels and promoting reproductive development in plants [44]. Overexpression of *MYB3R* transcription factor regulates cell cycle *cyclin* gene *OsCycB1;1*, playing a crucial role in G2/M phase transition during cold [22]. The expression of *OsCPT1*, a putative member of the DREB1 (dehydration-responsive element-binding factor 1)/ CBF pathway, leads to enhanced proline concentration, which ultimately increases tolerance to cold stress [45]. Using CRISPR-Cas9 gene editing, three genes were targeted in rice, *OsPIN5b* (a panicle length gene), *GS3* (a grain size gene), and *OsMYB30* (a cold tolerance gene) to improve cold tolerance. Nipponbare rice mutant of *Ospin5b*/gs3/ *Osmyb30*-4 resulted in improved cold stress tolerance and higher yield [23].

CONCLUSION

Flowering time/heading date in rice is controlled by two genetic pathways, *i.e.*, *OsGI–Hd1–Hd3a* and *Ehd1-Hd3a* pathways. The presence and absence of these alleles or reduced expression of genes associated with flowering, such as *HD1*, *Ehd1*, *RFT1*, and *Ghd7*, confer adaptabilities of rice varieties from temperate zone with long photoperiods to tropical zone with short photoperiods. Despite the natural variation, leading to the adaptability of rice to varied climate change, there is a need to accelerate high-quality rice production to address malnutrition and feed of ever-increasing population. Abiotic stress components, including the photoperiod, drought conditions, and temperature, affect the plant at several stages, but the most crucial stage is flowering. The flowering gene *Heading date-3a (Hd3a), Heading date 1 (Hd1), Rice Indeterminate 1 (RID1), Ghd7, Early heading date 1* (*Ehd1*), *Ehd2* and *Ehd3* are involved in flowering regulation. Fine-tuning rice heading dates by manipulating these genes can bring about some advantages, such as precise rotation of cropping systems with other profitable crops, enhancing more rice cropping per year, and developing superior varieties with high nutritional qualities. Adaptation of gene editing techniques such as TALENS, and CRISPR-Cas9 can speed up the process of functional characterization of flowering-associated genes. Also, combining gene editing technologies with breeding programs can reduce time in the process of crop improvement.

REFERENCE

[1]   Moonmoon S, Islam M. Effect of Drought Stress at Different Growth Stages on Yield and Yield Components of Six Rice (*Oryza sativa* L.) Genotypes. Fundamental and Applied Agriculture 2017; 2(3): 1.
[http://dx.doi.org/10.5455/faa.277118]

[2]   Yang X, Lu M, Wang Y, Wang Y, Liu Z, Chen S. Response mechanism of plants to drought stress. Horticulturae 2021; 7(3): 50.
[http://dx.doi.org/10.3390/horticulturae7030050]

[3]   Matsui T, Omasa K, Horie T. High temperature at flowering inhibits swelling of pollen grains, a driving force for thecae dehiscence in rice (*Oryza sativa* L.). Plant Prod Sci 2000; 3(4): 430-4.
[http://dx.doi.org/10.1626/pps.3.430]

[4]   Matsui T, Omasa K. Rice (*Oryza sativa* L.) cultivars tolerant to high temperature at flowering: anther characteristics. Ann Bot (Lond) 2002; 89(6): 683-7.
[http://dx.doi.org/10.1093/aob/mcf112] [PMID: 12102523]

[5]   Wang Y, Lu Y, Guo Z, Ding Y, Ding C. RICE CENTRORADIALIS 1, a TFL1-like Gene, Responses to Drought Stress and Regulates Rice Flowering Transition. Rice (N Y) 2020; 13(1): 70.
[http://dx.doi.org/10.1186/s12284-020-00430-3] [PMID: 32970268]

[6]   Ramu VS, Paramanantham A, Ramegowda V, Mohan-Raju B, Udayakumar M, Senthil-Kumar M. Transcriptome analysis of sunflower genotypes with contrasting oxidative stress tolerance reveals individual-And combined-biotic and abiotic stress tolerance mechanisms. PLoS One 2016; 11(6): e0157522.
[http://dx.doi.org/10.1371/journal.pone.0157522] [PMID: 27314499]

[7]   Babitha KC, Ramu SV, Nataraja KN, Sheshshayee MS, Udayakumar M. EcbZIP60, a basic leucine zipper transcription factor from Eleusine coracana L. improves abiotic stress tolerance in tobacco by activating unfolded protein response pathway. Mol Breed 2015; 35(9): 181.
[http://dx.doi.org/10.1007/s11032-015-0374-6]

[8]   Niranjan V, Uttarkar A, Dadi S, *et al.* Stress-Induced Detoxification Enzymes in Rice Have Broad Substrate Affinity. ACS Omega 2021; 6(4): 3399-410.
[http://dx.doi.org/10.1021/acsomega.0c05961] [PMID: 33553958]

[9]   Babitha K C, Ramu S V, Pruthvi V, Mahesh P, Nataraja K N, Udayakumar M. Co-expression of AtbHLH17 and AtWRKY28 confers resistance to abiotic stress in Arabidopsis Transgenic Res. 2013; 22: 327-41.

[10]   Babitha KC, Vemanna RS, Nataraja KN, Udayakumar M. Overexpression of EcbHLH57 transcription factor from Eleusine coracana L. in tobacco confers tolerance to salt, oxidative and drought stress. PLoS One 2015; 10(9): e0137098.
[http://dx.doi.org/10.1371/journal.pone.0137098] [PMID: 26366726]

[11] Xu Y, Chu C, Yao S. The impact of high-temperature stress on rice: Challenges and solutions. Crop J 2021; 9(5): 963-76.
[http://dx.doi.org/10.1016/j.cj.2021.02.011]

[12] Hirose F, Inagaki N, Hanada A, *et al.* Cryptochrome and phytochrome cooperatively but independently reduce active gibberellin content in rice seedlings under light irradiation. Plant Cell Physiol 2012; 53(9): 1570-82.
[http://dx.doi.org/10.1093/pcp/pcs097] [PMID: 22764280]

[13] Yang W, Gao M, Yin X, *et al.* Control of rice embryo development, shoot apical meristem maintenance, and grain yield by a novel cytochrome p450. Mol Plant 2013; 6(6): 1945-60.
[http://dx.doi.org/10.1093/mp/sst107] [PMID: 23775595]

[14] Pham VN, Kathare PK, Huq E. Phytochromes and phytochrome interacting factors. Plant Physiol 2018; 176(2): 1025-38.
[http://dx.doi.org/10.1104/pp.17.01384] [PMID: 29138351]

[15] Ghadirnezhad R, Fallah A. Temperature effect on yield and yield components of different rice cultivars in flowering stage. Int J Agron 2014; 1-4.
[http://dx.doi.org/10.1155/2014/846707]

[16] Jung JH, Domijan M, Klose C, *et al.* Phytochromes function as thermosensors in *Arabidopsis*. Science 2016; 354(6314): 886-9.
[http://dx.doi.org/10.1126/science.aaf6005] [PMID: 27789797]

[17] Shor E, Paik I, Kangisser S, Green R, Huq E. PHYTOCHROME INTERACTING FACTORS mediate metabolic control of the circadian system in Arabidopsis. New Phytol 2017; 215(1): 217-28.
[http://dx.doi.org/10.1111/nph.14579] [PMID: 28440582]

[18] Todaka D, Nakashima K, Maruyama K, *et al.* Rice phytochrome-interacting factor-like protein OsPIL1 functions as a key regulator of internode elongation and induces a morphological response to drought stress. Proc Natl Acad Sci USA 2012; 109(39): 15947-52.
[http://dx.doi.org/10.1073/pnas.1207324109] [PMID: 22984180]

[19] Ji X, Du Y, Li F, *et al.* The basic helix-loop-helix transcription factor, Os PIL 15, regulates grain size *via* directly targeting a purine permease gene *Os PUP 7* in rice. Plant Biotechnol J 2019; 17(8): 1527-37.
[http://dx.doi.org/10.1111/pbi.13075] [PMID: 30628157]

[20] Cai Q, Yuan Z, Chen M, *et al.* Jasmonic acid regulates spikelet development in rice. Nat Commun 2014; 5(1): 3476.
[http://dx.doi.org/10.1038/ncomms4476] [PMID: 24647160]

[21] Itoh JI, Nonomura KI, Ikeda K, *et al.* Rice plant development: from zygote to spikelet. Plant Cell Physiol 2005; 46(1): 23-47.
[http://dx.doi.org/10.1093/pcp/pci501] [PMID: 15659435]

[22] Ohmori Y, Tanaka W, Kojima M, Sakakibara H, Hirano HY. WUSCHEL-RELATED HOMEOBOX4 is involved in meristem maintenance and is negatively regulated by the CLE gene FCP1 in rice. Plant Cell 2013; 25(1): 229-41.
[http://dx.doi.org/10.1105/tpc.112.103432] [PMID: 23371950]

[23] Nagasaki H, Sato Y, Hong SK, Tagiri A, Kitano H, Yamamoto N. *et al.* A rice homeobox gene, OSH1, is expressed before organ differentiation in a specific region during early embryogenesis. Proc Natl Acad Sci U S A 93(8): 117-22.
[http://dx.doi.org/10.1073/pnas.93.15.8117]

[24] Fan T, Li X, Yang W, Xia K, Ouyang J, Zhang M. Rice osa-miR171c mediates phase change from vegetative to reproductive development and shoot apical meristem maintenance by repressing four OsHAM transcription factors. PLoS One 2015; 10(5): e0125833.
[http://dx.doi.org/10.1371/journal.pone.0125833] [PMID: 26023934]

[25]   Bäurle I, Dean C. The timing of developmental transitions in plants. Cell 2006; 125(4): 655-64.
       [http://dx.doi.org/10.1016/j.cell.2006.05.005] [PMID: 16713560]

[26]   Izawa T, Takahashi Y, Yano M. Comparative biology comes into bloom: genomic and genetic
       comparison of flowering pathways in rice and Arabidopsis. Curr Opin Plant Biol 2003; 6(2): 113-20.
       [http://dx.doi.org/10.1016/S1369-5266(03)00014-1] [PMID: 12667866]

[27]   Wu C, You C, Li C, *et al. RID1*, encoding a Cys2/His2-type zinc finger transcription factor, acts as a
       master switch from vegetative to floral development in rice. Proc Natl Acad Sci USA 2008; 105(35):
       12915-20.
       [http://dx.doi.org/10.1073/pnas.0806019105] [PMID: 18725639]

[28]   Turck F, Fornara F, Coupland G. Regulation and identity of florigen: FLOWERING LOCUS T moves
       center stage. Annu Rev Plant Biol 2008; 59(1): 573-94.
       [http://dx.doi.org/10.1146/annurev.arplant.59.032607.092755] [PMID: 18444908]

[29]   Zhao J, Chen H, Ren D, *et al.* Genetic interactions between diverged alleles of *Early heading date 1* (
       Ehd1) and *Heading date 3a* (*Hd3a*) / *RICE FLOWERING LOCUS T1* (*RFT 1*) control differential
       heading and contribute to regional adaptation in rice (*Oryza sativa*). New Phytol 2015; 208(3): 936-
       48.
       [http://dx.doi.org/10.1111/nph.13503] [PMID: 26096631]

[30]   Taoka K, Ohki I, Tsuji H, *et al.* 14-3-3 proteins act as intracellular receptors for rice Hd3a florigen.
       Nature 2011; 476(7360): 332-5.
       [http://dx.doi.org/10.1038/nature10272] [PMID: 21804566]

[31]   Kobayashi K, Yasuno N, Sato Y, *et al.* Inflorescence meristem identity in rice is specified by
       overlapping functions of three AP1/FUL-like MADS box genes and PAP2, a SEPALLATA MADS
       box gene. Plant Cell 2012; 24(5): 1848-59.
       [http://dx.doi.org/10.1105/tpc.112.097105] [PMID: 22570445]

[32]   Kim SL, Lee S, Kim HJ, Nam HG, An G. OsMADS51 is a short-day flowering promoter that
       functions upstream of Ehd1, OsMADS14, and Hd3a. Plant Physiol 2007; 145(4): 1484-94.
       [http://dx.doi.org/10.1104/pp.107.103291] [PMID: 17951465]

[33]   Izawa T, Mihara M, Suzuki Y, *et al.* Os-GIGANTEA confers robust diurnal rhythms on the global
       transcriptome of rice in the field. Plant Cell 2011; 23(5): 1741-55.
       [http://dx.doi.org/10.1105/tpc.111.083238] [PMID: 21571948]

[34]   Sun X, Zhang Z, Wu J, *et al.* The *Oryza sativa* Regulator HDR1 Associates with the Kinase OsK4 to
       Control Photoperiodic Flowering. PLoS Genet 2016; 12(3): e1005927.
       [http://dx.doi.org/10.1371/journal.pgen.1005927] [PMID: 26954091]

[35]   Wu W, Zheng XM, Lu G, *et al.* Association of functional nucleotide polymorphisms at *DTH2* with the
       northward expansion of rice cultivation in Asia. Proc Natl Acad Sci USA 2013; 110(8): 2775-80.
       [http://dx.doi.org/10.1073/pnas.1213962110] [PMID: 23388640]

[36]   Rao NN, Prasad K, Kumar PR, Vijayraghavan U. Distinct regulatory role for *RFL*, the rice *LFY*
       homolog, in determining flowering time and plant architecture. Proc Natl Acad Sci USA 2008; 105(9):
       3646-51.
       [http://dx.doi.org/10.1073/pnas.0709059105] [PMID: 18305171]

[37]   Ryu CH, Lee S, Cho LH, *et al. OsMADS50* and *OsMADS56* function antagonistically in regulating
       long day (LD)-dependent flowering in rice. Plant Cell Environ 2009; 32(10): 1412-27.
       [http://dx.doi.org/10.1111/j.1365-3040.2009.02008.x] [PMID: 19558411]

[38]   Doi K, Izawa T, Fuse T, *et al. Ehd1*, a B-type response regulator in rice, confers short-day promotion
       of flowering and controls *FT-like* gene expression independently of *Hd1*. Genes Dev 2004; 18(8): 926-
       36.
       [http://dx.doi.org/10.1101/gad.1189604] [PMID: 15078816]

[39]   Izawa T. Adaptation of flowering-time by natural and artificial selection in Arabidopsis and rice. J Exp

Bot 2007; 58(12): 3091-7.
[http://dx.doi.org/10.1093/jxb/erm159] [PMID: 17693414]

[40]   Kim SR, Torollo G, Yoon MR, *et al.* Loss-of-function alleles of heading date 1 (Hd1) are associated with adaptation of temperate japonica rice plants to the tropical region. Front Plant Sci 2018; 9: 1827.
[http://dx.doi.org/10.3389/fpls.2018.01827] [PMID: 30619400]

[41]   Xue W, Xing Y, Weng X, *et al.* Natural variation in Ghd7 is an important regulator of heading date and yield potential in rice. Nat Genet 2008; 40(6): 761-7.
[http://dx.doi.org/10.1038/ng.143] [PMID: 18454147]

[42]   Yin M, Ma H, Wang M, *et al.* Transcriptome analysis of flowering regulation by sowing date in Japonica Rice (*Oryza sativa* L.). Sci Rep 2021; 11(1): 15026.
[http://dx.doi.org/10.1038/s41598-021-94552-3] [PMID: 34294838]

[43]   Liu X, Liu H, Zhang Y, He M, Li R, Meng W, *et al.* Fine-tuning Flowering Time *via* Genome Editing of Upstream Open Reading Frames of Heading Date 2 in Rice. 2021; 82: 239-53.

[44]   Guo C, Ge X, Ma H. The rice OsDIL gene plays a role in drought tolerance at vegetative and reproductive stages. Plant Mol Biol 2013; 82(3): 239-53.
[http://dx.doi.org/10.1007/s11103-013-0057-9] [PMID: 23686450]

[45]   Ma Q, Dai X, Xu Y, *et al.* Enhanced tolerance to chilling stress in OsMYB3R-2 transgenic rice is mediated by alteration in cell cycle and ectopic expression of stress genes. Plant Physiol 2009; 150(1): 244-56.
[http://dx.doi.org/10.1104/pp.108.133454] [PMID: 19279197]

CHAPTER 2

A Peep into the Tolerance Mechanism and the Sugar Beet Response to Salt Stress

Varucha Misra[1,*] and **Ashutosh Kumar Mall**[1]

[1] *ICAR-Indian Institute of Sugarcane Research Lucknow-226 002, (U.P.), India*

Abstract: Salt stress is one of the main environmental stresses occurring all over the globe. Soil salinity is a serious issue in arid and semi-arid areas, causing significant ecological disruption. Excess salts in the soil have an impact on plant nutrient intake and osmotic equilibrium, causing osmotic and ionic stress. Complex physiological features, metabolic pathways, enzyme synthesis, suitable solutes, metabolites, and molecular or genetic networks all play a role in plant adaptation or tolerance to salinity stress. Sugar beet is a well-known crop in terms of salt tolerance and for reclaiming such soils, even for the growth of other crops. Natural endowments, accumulation of organic solutes, sodium potassium ions accumulation in vacuoles, and osmotic tolerance potential are some of the key mechanisms involved in providing tolerance to sugar beet. A greater understanding of sugar beet tolerance and response to salt stress will open up new avenues for increasing crop performance in these conditions. The mechanisms involved in sugar beet adaptation to salt stress conditions, as well as the response to such conditions, are discussed in this chapter.

Keywords: Osmotic adjustment, Mechanism, SOS Pathway, Sugar beet, Tolerance.

1. INTRODUCTION

Salinity is a severe challenge to modern agriculture, causing crop growth and development to be hampered. Excess salt levels harm around 960 million hectares of fertile land worldwide [1 - 3]. In the current situation, more than 20% of well-irrigated lands around the world are vulnerable to soil saline conditions. Due to the excessive use of water for crop irrigation and rapid barrenness (due to climate change issues), the problem of salt stress has grown [4]. The increasing soil salinity is majorly due to the global variation in temperature, poor irrigation practices, and incorrect fertilizer use. All of these have enhanced the negative impacts of salt increment in soils resulting in an annual salinity increase of 1-2 percent [5]. Munns [6] and Munns [7] described a two-phase growth reaction of

---
[*] **Corresponding author Varucha Misra :** ICAR-Indian Institute of Sugarcane Research Lucknow 226 002, (U.P.), India; Email: misra.varucha@gmail.com

Jen-Tsung Chen (Ed.)

plants towards salt stress. In the first phase, the effect of salt stress has contributed to osmotic alterations occurring exterior to the salt-affected cells. This, in turn, causes a lower rate of water absorption. The second phase is denoted by salt accumulation in leaves which results in enzymatic activity hindrance and activation of ionic and oxidative stress [8].

Sugar beet, thanks to its sea beet ancestor, has a high salt tolerance capacity [9, 10] of 9.5 m mhos/cm to salt stress [11, 12]. This salt tolerance ability is exceptionally strong when compared to other crops, such as paddy (6.0), sugarcane (4.4), and wheat (2.4). It's a halophyte that scavenges sodium salts in salty soils, removing roughly 500 kg of sodium salts per hectare per season [11]. Several cultivars have obtained salt tolerance traits from their ancestor. Yang *et al.* [9] stated that this crop can withstand 500 mM sodium chloride (NaCl) for seven days without dying. Sugar beet was found to be able to grow in soil containing 85–140 mM salt [13]. According to Peng *et al.* [14], sugar beets grew better in 3 mM NaCl than in 0 mM NaCl. Besides, the production quality had even been reported to be ineffective when the electrical conductivity (EC) of the soil reached 7.0 dS m^{-1} (67 mM NaCl) [15]. However, when EC exceeds the 7.0 dS m^{-1} level, this crop shows reduced yield [16, 17]. Khandil *et al.* [18] reported this crop as an economically farmed crop in saline soils and also helped in reclaiming such soils. Sugar beet is now considered an ideal crop model for understanding salt tolerance mechanisms, thanks to the completion of its genome sequencing [19].

1.1. Characteristics of Halophytes for Salt Stress Condition

Halophytes have a built-in ability to thrive in saline environments. Chenopodiaceous plants have more mitochondria than other plants because they require more energy in high-salt environments [20]. Additionally, these plants aid in the accumulation of sodium and chlorine ions in their cytoplasm. This allows the chloroplasts to withstand even the saline conditions are at extreme [21]. These plants also contain special glands on their leaves that aid in the elimination of salt. These glands prevent salt from reaching the sprout *via* the surface of the leaves [22]. Furthermore, the presence of hydathodes in these plants aid in the excretion of excessive salts. With less stomatal conductance and more transpiration water loss, this happens.

1.2. Salt Stress Tolerance Mechanism in Sugar Beet

The tolerance mechanism in sugar beet for salt stress has been investigated in a number of studies [23, 24]. Osmotic tolerance, ion exclusion, and tissue tolerance are the three identified mechanisms behind the tolerance potential of this plant for such conditions. Osmotic tolerance/ adjustment potential has been prominently seen in this crop with the accrual of sodium and chloride ions in their vacuoles and cytoplasm [25, 26]. It is an essential regulatory mechanism for the crop to adapt to salt conditions [27]. An osmotic gradient is created when salt components from the cytoplasm accumulate in the vacuole. The surge in the solute synthesis that happens in the cytoplasm controls this gradient. Osmotic adjustment is the process of increasing solute production [28]. This is an adaptive approach for plants that have been exposed to a saltwater environment. Various suitable solutes have been discovered. Compatible solutes have features such as increased solubility, low polar charge, and a broad hydration shell [29]. The various features of compatible solutes contribute to cell turgor maintenance and even cause consistency in enzyme structure in the cytoplasm. As a result, the solutes are protected by inorganic ions [30]. Flowers [31] found the absorbing and accumulating capability of sodium ions in the sugar beet leaves for osmotic regulation/adaptation under saline conditions. Flowers and Colmer [32] found that plants grown under salt-stress conditions possess high rates of organic solutes. The production of organic solutes under such a condition causes the plant to maintain its cell turgidity and enzymatic conformation. Proline, in this respect, plays an essential part. It is involved in the protection of the cytomembrane system and sustaining the enzyme structure [33]. Ashraf and Fatima [34] stated that proline production varies with tolerance potential for salt stress. Proline and soluble sugars have long been recognized for their role in osmotic adjustment in plants under salt stress [35].

Zeng *et al.* [36] found that salt-tolerant halophytes had a substantial increase in betaine contents. Wang *et al.* [37] showed that the accumulation of amino acids in salt-stressed sugar beet had much higher rates than proline/choline levels. Wang *et al.* [38] depicted that the higher the organic acid in salt-stressed plants higher the salt tolerance potential. They further stated that organic acid accumulation causes the plant to cope with abiotic stresses. Iqbal *et al.* [39] revealed that protein structure and cell membrane stability are sustained by salt-tolerant sugar beet plants in the osmotic adjustment defense mechanism. This helps in lowering the osmotic potential of cells. Ninety-five redox proteins have been identified in this crop under salt stress conditions depicting its involvement in metabolic activities as well as several other activities related to defense mechanisms for such an environment [40]. Mansour and Ali [41] have shown that all these responses help in giving energy to plants for coping/adapting to the salt stress condition.

Sugar beet has long been known to have a salt stress tolerance capacity. Yang *et al.* [9] revealed that this plant can withstand the salt stress condition of up to 500 mM NaCl for one week and that too without losing its viability. According to Gupta [42], the tolerance power of sugar beet was 9.5 Dsm^{-1}. The absorption of salt by the plant from the soil to the shoots has been documented in this trait. This aids in the preservation of osmotic equilibrium, but it can also induce toxicity and nutritional imbalance in plants [15, 43]. Koyro and Huchzermeyer [23] have shown that sugar beet has the potential to store soluble sugars in its roots which provides support in the regulation of osmotic potential when salt stress occurs. Eisa and Ali [24] revealed the purpose behind the prominent impact of salt stress in germination and early growth stages of this crop. This is so as the rate of photosynthesis becomes significantly reduced, resulting in little production of sugars in roots. Ayars and Schoneman [44] found this crop to be well-grown under moderate salt stress conditions.

The accrual of sodium and potassium ions in the vacuolar and cytoplasmic regions of the cells is another strategy that the beet crop adopts for showing tolerance for such situations [26, 45]. The accumulation and absorption of sodium ions by this plant under salt stress conditions help in regulation and adaptation to osmotic potential with soil [31]. Glenn and Brown [46] showed the precise mechanism for controlling the uptake of Na^+ and Cl^- ions for water intake. Flowers and Colmer [32] proved this mechanism as the reason for the halophytes plant's tolerance capability, including sugar beet. Besides ion uptake, cellular compartmentalization and organic solute formation also work together in providing tolerance to sugar beet under salt stress conditions. Na^+/H^+ antiporters play an important function in providing salt tolerance to plants. These antiporters help in expelling the sodium ions from the cytoplasm [47]. Parallelly, the H^+ gradient created by H^+-ATPases and H^+-PPases initiates sodium ions to move into the vacuole [48]. This is the reason why there is a necessity for hydrogen ion pumps in the cytoplasm [49].

Potassium ions also play an important role in sugar beet under salt stress conditions. As per Lindhauer *et al.* [43], reduced levels of K^+ ions have been reported due to the osmotic regulation in this salt stress condition. Osmotic adjustment through sodium, potassium, and chloride ions increased the growth potential of sugar beet under salt stress conditions [26]. Lindhauer *et al.* [42] found that ions of potassium, sodium, and magnesium may help in obtaining potential osmotic adjustment in leaves of sugar beet. This has made this crop a choice for user lands where cultivation of other crops fails. Shrivastava *et al.* [11] showed its potential yield in salt areas of Sunderbans after paddy cultivation. This indicated its lucrative capability in severe salt areas.

1.3. Salt Overly Sensitive (SOS) Pathway for Salt Tolerance

This pathway involves the use of three parts, *viz.*, SOS 1, SOS 2, and SOS 3. These three parts function together in regulating the Na^+ concentrations in the cytosol. Thus, the pathway plays an important function in controlling/regulating ion homeostasis, where Na^+ is transported outside the cell through these signals. It is considered an essential signaling pathway for a crop to stimulate during salt stress conditions and serves as a defense mechanism for the crop. Research on the SOS pathway in halophytes on molecular and physiological levels is limited [22, 50, 51]. A calcium signal is stimulated when Na^+ ions concentration is increased, and this, in turn, causes activation of the SOS pathway. Geng *et al.* [52] revealed that calmodulin-like protein (CML7 gene) expression was increased in salt-tolerant sugar beet cultivar. This also showed that there is the involvement of calcium ion sensors in the regulation of salt stress tolerance in sugar beet. In the cell membrane, the calcium-binding proteins SOS 3 and SOS 2 form a complex that causes activation of SOS 1 antiporter [53]. This pathway also provides tolerance to crops by a cellular process dependent on Na^+ efflux, which returns sodium ions either to the soil or to the apoplast Fig (**1**).

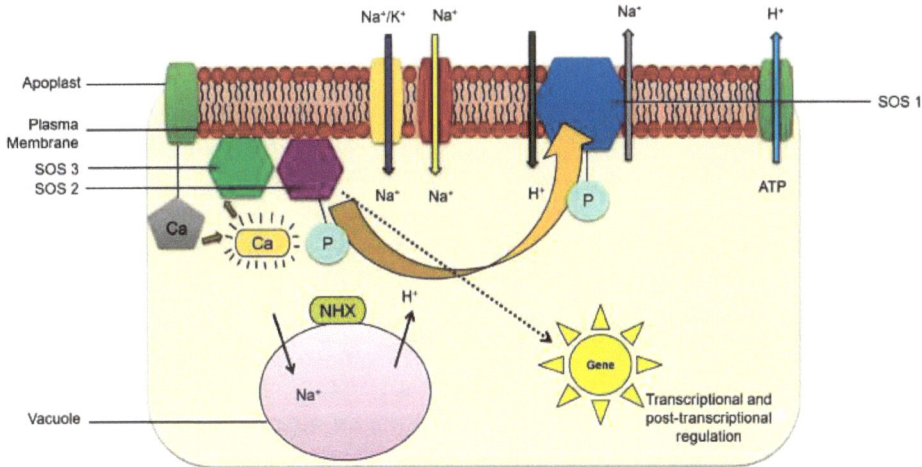

Fig. (1). The SOS signaling pathway in generalized plants under salt stress conditions.

1.4. The Response of Sugar Beet under Salt Stress

Production and productivity are both affected by salt concentration in any crop. Variations in salt concentration have a strong impact on the growth and growth-related traits of sugar beet. The uptake of water and nutrients essential for crop growth is impeded by the variation of salt concentration in any crop [54]. In sugar beet, salt stress causes different effects on the crop stage at which it strikes Table (**1**). Abbas *et al.* [55] found that the tolerant nature of this crop was seen during vegetative growth, while the sensitive stage has been reported as germination, seedling, and emergence [35, 56]. Ayers and Hayward [57] showed that at the germination phase, when EC of saline soil is at 6 mmho $cm^{-1,}$ then a reduction of

50% germination rate was seen. Ion toxicity, nutritional imbalance, photosynthetic efficiency, and other related processes have a repressed effect on sugar beet under salt stress conditions [58]. Besides, sugar beet, as a response to salt stress conditions, also shows physio-biochemical adaptive traits Table (**2**).

Table 1. Sugar beet response at various crop stages under saline environment.

Stage	Response	Impact	References
Seedling emergence	Reduction	Lowers the number of seedlings	[59, 60]
Germination	Poor germination	Ion toxicity, Restriction/hindrance in cell division process, and rise in the growth point.	[26, 61]
Vegetative growth	Reduced shoot content	Reduction in fresh weight	[62]
Flowering	Early flowering	Increase in root shoot ratio, increase in leaf size, Dry matter reduced	[61]

The first and foremost striking impact of salt stress in sugar beet under growing stages has been seen in roots. An increase in salt concentrations near the root zone of sugar beet causes a strong impact on growth as the cells become deficient with water, have nutritional disorders, and increase toxic ions [63]. A decrease in root length in beet plants is a common trait under salt stress conditions. Ghoulam *et al.* [26] found variation in two genotypes under salt stress conditions based on sensitivity and tolerance to such situations. H 30917 genotype had the longest root length while H 30973 genotype did not. Sugar beet leaves are another portion that gets affected by salt stress conditions. Studies showed that the fresh and dry weight of roots and shoots were lowered in salt stress conditions, but both these traits were linked with the severity status of the salt stress condition [64 - 66]. In general, plant effect towards salt stress shows changes in physio-biochemical processes. These processes include ion balance, reactive oxygen species production, and regulation of osmotic potential. A large number of genes regulate all these processes by stimulating expression. Li *et al.* [67] identified four ncRNAs in roots playing an important role in salt stress conditions. They further found that these stress-responsive genes were relatively higher in roots than in leaves. The mobilization capability of roots under such conditions has been reported to be wider. Root cells change numerous processes to shift resources from growth and development to survival, according to a functional study of Differentially expressed genes (DEGs) in roots. The upregulation of sucrose synthases in roots has also been shown [67].

Ghoulam *et al.* [26] revealed that the beets leaf area is strongly affected by salt stress as such condition causes a lowering in the leaves' expansion rate [68]. Witkwski and Lamont [69] found the reason for these low-expanded beet leaves in such a situation. Relative water content in leaves is another trait that gets reduced [26]. Leaf thickness is the additional parameter that gets altered. The increase in the thickness of beet leaves has been revealed as an associated response in sugar beets in such an environment [12]. Stevanato *et al.* [70] stated that when the salt concentration is low (70 mM NaCl) then, the aerial portion of the plants has good growth potential. This depicts that the aerial portion has a higher tolerance capability in seedlings. Furthermore, the sodium and chloride ions were high in petioles compared to roots/leaves in salt stress conditions. The accumulation of sodium ions in leaf tissues under salt stress conditions may help in the protection of the photosynthetic system from the ion toxicity generated by salt stress conditions [71]. Besides, Stevanato *et al.* [70] revealed that older beet leaves accumulate more salt ions. This, in turn, protects the younger leaves from the effect of salt stress conditions. Furthermore, Hajiboland *et al.* [72] revealed that under salt-stress conditions, beet leaves became succulent, and these leaves were thicker. This helps in storing/accumulating higher water per unit leaf area. This trait is a morphological adaptive measure of sugar beet under salt stress conditions. Hampe and Marschner [73] stated that the combination of succulent leaves with more leaf area had been an important trait for sugar beet under salt stress conditions.

Table 2. Physio-biochemical response of sugar beet in salt stress condition.

Physiological Response	References	Biochemical Response	References
Reduction in chlorophyll content	[74]	Compatible solutes accumulation	[8]
Photosynthetic efficiency and rate were lowered	[8]	Oxidative stress increased	[72]
Stomatal conductance reduced	[8]	Reactive oxygen species accumulation	[72]
Germination rate and seedling growth drastically lowered	[8]	Antioxidant enzymatic activities altered (Superoxide dismutase; Catalase; Ascorbate peroxidase)	[78]
Root length shortened	[75]	A rise in proline content	[79]
Root weight reduced	[75]		
Leaf temperature increased	[76]		
Leaf succulence increased	[72]	Rise in glycine betaine	[45]
Sodium-ions percentage in leaves varies	[72]		
The H^+-ATPase activity of plasma membrane reduced in tolerant varieties	[77]		

CONCLUSION

Salt stress is a common and important problem worldwide, affecting most crops' production and yield. Roughly 960 million hectares of arable land have been covered under saline conditions worldwide. Salt stress is responsible for ionic imbalance and osmotic stress, along with the generation of other multiple stresses. All these combine to cause a significant effect on different stages of the crop. Sugar beet has a high tolerance potential for salt stress conditions. It is amongst those crops that survive well under such conditions and also reclaim the soil (500 kg Na salts ha^{-1} per season) for the growth of other crops. It has a tolerance potential of 9.5 m mhos/cm in such conditions. Besides, there have been several other adaptive responses of this crop towards salt stress, like osmoregulation, osmotic stress adjustment, osmolyte production, *etc.*

New sugar beet cultivars with steady and high yields are desperately needed for salt stress conditions. In sugar beet, several salt tolerance characteristics and signaling pathways have been studied over the last two decades. Candidate genes have been reported to be a participant in osmolyte biosynthesis in sugar beet. Limited studies have been focused on molecular markers and quantitative trait loci related to the salt tolerance potential of sugar beet. There is a need for further investigation on this aspect, with more emphasis on signal transduction pathways under such conditions. Several fields about sugar beet tolerance still need to be researched, like cross talk, developmental regulation, *etc.* Sugar beet transcriptomics and proteomics research are being done, yet there is a need for elaborative studies on this aspect to achieve a clear picture of the salt tolerance potential of this crop.

REFERENCES

[1]   Munns R, Tester M. Mechanisms of salinity tolerance. Annu Rev Plant Biol 2008; 59(1): 651-81.
[http://dx.doi.org/10.1146/annurev.arplant.59.032607.092911] [PMID: 18444910]

[2]   Deinlein U, Stephan AB, Horie T, Luo W, Xu G, Schroeder JI. Plant salt-tolerance mechanisms. Trends Plant Sci 2014; 19: 371-9.
[http://dx.doi.org/10.1016/j.tplants.2014.02.001]

[3]   Jha UC, Bohra A, Jha R, Parida SK. Salinity stress response and 'omics' approaches for improving salinity stress tolerance in major grain legumes. Plant Cell Rep 2019; 38(3): 255-77.
[http://dx.doi.org/10.1007/s00299-019-02374-5] [PMID: 30637478]

[4]   Machado R, Serralheiro R. Soil salinity: effect on vegetable crop growth. Management practices to prevent and mitigate soil salinization. Horticulturae 2017; 3(2): 30.
[http://dx.doi.org/10.3390/horticulturae3020030]

[5]   Zhu JK. Abiotic stress signaling and responses in plants. Cell 2016; 167(2): 313-24.
[http://dx.doi.org/10.1016/j.cell.2016.08.029] [PMID: 27716505]

[6]   Munns R. Comparative physiology of salt and water stress. Plant Cell Environ 2002; 25(2): 239-50.
[http://dx.doi.org/10.1046/j.0016-8025.2001.00808.x] [PMID: 11841667]

[7]   Munns R. Genes and salt tolerance: bringing them together. New Phytol 2005; 167(3): 645-63.
[http://dx.doi.org/10.1111/j.1469-8137.2005.01487.x] [PMID: 16101905]

[8]   Skorupa M, Szczepanek J, Mazur J, Domagalski K, Tretyn A, Tyburski J. Salt stress and salt shock differently affect DNA methylation in salt-responsive genes in sugar beet and its wild, halophytic ancestor. PLoS One 2021; 16(5): e0251675.
[http://dx.doi.org/10.1371/journal.pone.0251675] [PMID: 34043649]

[9]   Yang L, Ma C, Wang L, Chen S, Li H. Salt stress induced proteome and transcriptome changes in sugar beet monosomic addition line M14. J Plant Physiol 2012; 169(9): 839-50.
[http://dx.doi.org/10.1016/j.jplph.2012.01.023] [PMID: 22498239]

[10]   Wedeking R, Mahlein AK, Steiner U, Oerke EC, Goldbach HE, Wimmer MA. Osmotic adjustment of young sugar beets (*Beta vulgaris*) under progressive drought stress and subsequent rewatering assessed by metabolite analysis and infrared thermography. Funct Plant Biol 2017; 44(1): 119-33.
[http://dx.doi.org/10.1071/FP16112] [PMID: 32480551]

[11]   Shrivastava AK, Sawnani A, Shukla SP, Solomon S. Unique features of sugarbeet and its comparison with sugarcane Souvenir of IISR Industry interface on Research and development initiatives for sugar beet in India. ICAR-IISR Lucknow 2013; pp. 36-9.

[12]   Misra V, Mall AK, Pathak AD. Sugar beet: A sustainable crop for saline environment.Agronomic Crops. Singapore: Springer 2020; pp. 49-61.
[http://dx.doi.org/10.1007/978-981-15-0025-1_4]

[13]   Li W, Wang R, Wang W, *et al.* Effect of NaCl stress on sugar beet growth. Zhongguo Tangliao 2007; 2: 17-9.
[http://dx.doi.org/10.3969/j.issn.1007-2624.2007.02.006]

[14]   Peng C, Geng G, Yu L, *et al.* Effect of different Na+ concentrations on growth and physiological traits of sugar beet. Zhiwu Yingyang Yu Feiliao Xuebao 2014; 20: 459-65.
[http://dx.doi.org/10.11674/zwyf.2014.0223]

[15]   Marschner H. Mineral nutrition of higher plants. 2nd ed., London, UK: Academic Press 1995.

[16]   Rhoades JD, Loveday J. Salinity in irrigated agriculture.Irrigation of Agricultural Crops Agronomy Monograph no 30. Madison: American Society of Agronomy Inc. 1990; pp. 1089-142.

[17]   Maas E. Crop salt tolerance.Agricultural salinity and assessment ASCE Manual on Engineering Practice. New York: ASCE 1990; p. 71.

[18]   Kandil AA, Sharief AE, Abido WAE, Awed AM. Effect of gibberellic acid on germination behavior of sugar Beet cultivars under salt stress conditions of Egypt. Sugar Tech 2014; 16(2): 211-21.
[http://dx.doi.org/10.1007/s12355-013-0252-7]

[19]   Dohm JC, Minoche AE, Holtgräwe D, *et al.* The genome of the recently domesticated crop plant sugar beet (Beta vulgaris). Nature 2014; 505(7484): 546-9.
[http://dx.doi.org/10.1038/nature12817] [PMID: 24352233]

[20]   Slama I, Abdelly C, Bouchereau A, Flowers T, Savouré A. Diversity, distribution and roles of osmoprotective compounds accumulated in halophytes under abiotic stress. Ann Bot (Lond) 2015; 115(3): 433-47.
[http://dx.doi.org/10.1093/aob/mcu239] [PMID: 25564467]

[21]   Sévin DC, Stählin JN, Pollak GR, Kuehne A, Sauer U. Global metabolic responses to salt stress in fifteen species. PLoS One 2016; 11(2): e0148888.
[http://dx.doi.org/10.1371/journal.pone.0148888] [PMID: 26848578]

[22]   Shabala S. Learning from halophytes: physiological basis and strategies to improve abiotic stress tolerance in crops. Ann Bot (Lond) 2013; 112(7): 1209-21.
[http://dx.doi.org/10.1093/aob/mct205] [PMID: 24085482]

[23]   Koyro HW, Huchzermeyer B. The physiological response of *Beta vulgaris* subsp. *maritima* to

seawater irrigation. Proceedings of International Conference on water management, salinity and pollution control towards sustainable irrigation in the Mediterranean region. Bari, Italy. 1997; pp. 29-49.

[24]   Eisa S, Ali S. Biochemical, Physiological and Morphological responses of sugar beet to salinization. J Plant Production 2005; 30(9): 5231-42.
[http://dx.doi.org/10.21608/jpp.2005.237478]

[25]   Subbarao GV, Wheeler RM, Levine LH, Stutte GW. Glycine betaine accumulation, ionic and water relations of red-beet at contrasting levels of sodium supply. J Plant Physiol 2001; 158(6): 767-76.
[http://dx.doi.org/10.1078/0176-1617-00309] [PMID: 12033231]

[26]   Ghoulam C, Foursy A, Fares K. Effects of salt stress on growth, inorganic ions and proline accumulation in relation to osmotic adjustment in five sugar beet cultivars. Environ Exp Bot 2002; 47(1): 39-50.
[http://dx.doi.org/10.1016/S0098-8472(01)00109-5]

[27]   Volkmar KM, Hu Y, Steppuhn H. Physiological responses of plants to salinity: A review. Can J Plant Sci 1998; 78(1): 19-27.
[http://dx.doi.org/10.4141/P97-020]

[28]   McCue K, Hanson A. Drought and salt tolerance: towards understanding and application. Trends Biotechnol 1990; 8: 358-62.
[http://dx.doi.org/10.1016/0167-7799(90)90225-M]

[29]   Paleg LG, Stewart GR, Starr R. The effect of compatible solutes on proteins. Plant Soil 1985; 89(1-3): 83-94.
[http://dx.doi.org/10.1007/BF02182235]

[30]   Sommer C, Thonke B, Popp M. The compatibility of D-pinitol and 1D-1-O-methyl-mucoinositol with malate dehydrogenase activity. Bot Acta 1990; 103(3): 270-3.
[http://dx.doi.org/10.1111/j.1438-8677.1990.tb00160.x]

[31]   Flowers TJ. Chloride as a nutrient and as an osmoticum. Advances in Plant Nutrition 1988; 3: 55-78.

[32]   Flowers TJ, Colmer TD. Salinity tolerance in halophytes. New Phytol 2008; 179(4): 945-63.
[http://dx.doi.org/10.1111/j.1469-8137.2008.02531.x] [PMID: 18565144]

[33]   Hong Z, Lakkineni K, Zhang Z, Verma DPS. Removal of feedback inhibition of delta(1)-pyrroline-5-carboxylate synthetase results in increased proline accumulation and protection of plants from osmotic stress. Plant Physiol 2000; 122(4): 1129-36.
[http://dx.doi.org/10.1104/pp.122.4.1129] [PMID: 10759508]

[34]   Ashraf M, Fatima H. Responses of some salt-tolerant and salt-sensitive lines of safflower (*Carthamus tinctorius* L.). Acta Physiol Plant 1995; 17: 61-71.

[35]   Farkhondeh R, Nabizadeh E, Jalilnezhad N. Effect of salinity stress on proline content, membrane stability and water relations in two sugar beet cultivars. Int J Agric Sci 2012; 2(5): 385-92.

[36]   Zeng Y, Li L, Yang R, Yi X, Zhang B. Contribution and distribution of inorganic ions and organic compounds to the osmotic adjustment in *Halostachys caspica* response to salt stress. Sci Rep 2015; 5: 136-9.
[http://dx.doi.org/10.1038/srep13639]

[37]   Wang Y, Stevanato P, Yu L, *et al.* The physiological and metabolic changes in sugar beet seedlings under different levels of salt stress. J Plant Res 2017; 130(6): 1079-93.
[http://dx.doi.org/10.1007/s10265-017-0964-y] [PMID: 28711996]

[38]   Wang M, Li P, Li C, *et al.* SiLEA14, a novel atypical LEA protein, confers abiotic stress resistance in foxtail millet. BMC Plant Biol 2014; 14(1): 290.
[http://dx.doi.org/10.1186/s12870-014-0290-7] [PMID: 25404037]

[39]   Iqbal N, Umar S, Khan NA, Khan MIR. A new perspective of phytohormones in salinity tolerance:

Regulation of proline metabolism. Environ Exp Bot 2014; 100: 34-42.
[http://dx.doi.org/10.1016/j.envexpbot.2013.12.006]

[40]   Liu H, Du X, Zhang J, *et al.* Quantitative redox proteomics revealed molecular mechanisms of salt tolerance in the roots of sugar beet monomeric addition line M14. Bot Stud (Taipei, Taiwan) 2022; 63(1): 5.
[http://dx.doi.org/10.1186/s40529-022-00337-w]

[41]   Mansour MMF, Ali EF. Evaluation of proline functions in saline conditions. Phytochemistry 2017; 140: 52-68.
[http://dx.doi.org/10.1016/j.phytochem.2017.04.016] [PMID: 28458142]

[42]   Gupta DK. Prospect of sugar beet based commercially viable industries in the backward coastal saline tract of the country.Transferable Technology for Rural Development. New Delhi: Associated Publishing Company 1985; pp. 124-33.

[43]   Lindhauer MG, Haeder HE, Beringer H. Osmotic potentials and solute concentrations in sugar beet plants cultivated with varying potassium/sodium ratios. Z Pflanzenernährung Bodenkd 1990; 153(1): 25-32.
[http://dx.doi.org/10.1002/jpln.19901530107]

[44]   Ayars JE, Schoneman RA. Irrigating field crops in the presence of saline ground water. Guangai Paishui Xuebao 2006; 55(3): 265-79.
[http://dx.doi.org/10.1002/ird.258]

[45]   Subbarao GV, Wheeler RM, Levine LH, Stutte GW. Glycine betaine accumulation, ionic and water relations of red-beet at contrasting levels of sodium supply. J Plant Physiol 2001; 158(6): 767-76.
[http://dx.doi.org/10.1078/0176-1617-00309] [PMID: 12033231]

[46]   Glenn EP, Brown JJ, Blumwald E. Salt tolerance and crop potential of halophytes. Crit Rev Plant Sci 1999; 18(2): 227-55.
[http://dx.doi.org/10.1080/07352689991309207]

[47]   Tester M, Davenport R. Na+ tolerance and Na+ transport in higher plants. Ann Bot (Lond) 2003; 91(5): 503-27.
[http://dx.doi.org/10.1093/aob/mcg058] [PMID: 12646496]

[48]   Apse MP, Aharon GS, Snedden WA, Blumwald E. Salt tolerance conferred by overexpression of a vacuolar Na+/H+ antiport in *Arabidopsis.* Science 1999; 285(5431): 1256-8.
[http://dx.doi.org/10.1126/science.285.5431.1256] [PMID: 10455050]

[49]   Chinnusamy V, Jagendorf A, Zhu JK. Understanding and improving salt tolerance in plants. Crop Sci 2005; 45(2): 437-48.
[http://dx.doi.org/10.2135/cropsci2005.0437]

[50]   Quintero FJ, Ohta M, Shi H, Zhu JK, Pardo JM. Reconstitution in yeast of the *Arabidopsis* SOS signaling pathway for Na$^+$ homeostasis. Proc Natl Acad Sci USA 2002; 99(13): 9061-6.
[http://dx.doi.org/10.1073/pnas.132092099] [PMID: 12070350]

[51]   Oh DH, Dassanayake M, Haas JS, *et al.* Genome structures and halophyte-specific gene expression of the extremophile *Thellungiella parvula* in comparison with *Thellungiella salsuginea* (*Thellungiella halophila*) and *Arabidopsis.* Plant Physiol 2010; 154(3): 1040-52.
[http://dx.doi.org/10.1104/pp.110.163923] [PMID: 20833729]

[52]   Geng G, Lv C, Stevanato P, *et al.* Transcriptome analysis of salt sensitive and tolerant genotypes reveals salt tolerance metabolic pathways in sugar beet. Int J Mol Sci 2019; 20(23): 5910.
[http://dx.doi.org/10.3390/ijms20235910] [PMID: 31775274]

[53]   Guo KM, Babourina O, Rengel Z. Na$^+$/H$^+$ antiporter activity of the *SOS1* gene: lifetime imaging analysis and electrophysiological studies on Arabidopsis seedlings. Physiol Plant 2009; 137(2): 155-65.
[http://dx.doi.org/10.1111/j.1399-3054.2009.01274.x] [PMID: 19758408]

[54] David F. Salt accumulation processes. Fargo, ND: North Dakota State University 2007; p. 58105.

[55] Abbas F, Mohanna A, Al-Lahham G. AL-Jbawi E, AL-Jasem Z. Evaluating the response of some sugar beet (*Beta vulgaris* L.) genotypes under saline water irrigation conditions. Arab J Arid Eniron 2011; 4(1): 93-105.

[56] Kaffka S, Hembree K. The effects of saline soil, irrigation, and seed treatments on sugar beet stand establishment. J Sugar Beet Res 2004; 41(3): 61-72.
[http://dx.doi.org/10.5274/jsbr.41.3.61]

[57] Ayers AD, Hayward HE. A method for measuring the effect of soil salinity on seed germination with observations on several crop plants. Soil Science Society Proceedings 2006; 13: 221.

[58] Khan AH, Ashraf MY, Naqvi SSM, Khanzada B, Ali M. Growth and ion and solute contents of sorghum grown under NaCl and Na_2SO_4 salinity stress. Acta Physiol Plant 1995; 17: 261-8.

[59] Francois LE, Goodin JR. Interaction of temperature and salinity on sugar beet germination. Agron J 1972; 64(3): 272-3.
[http://dx.doi.org/10.2134/agronj1972.00021962006400030004x]

[60] Beatty KD, Ehlig CF. A technique for testing and selecting for salt tolerance in sugar beet. J Sugar Beet Res 1973; 17(4): 295-9.
[http://dx.doi.org/10.5274/jsbr.17.4.295]

[61] Maghsoudi MA, Maghsoudi K. Salt Stress effects on respiration and growth of germinated seeds of different wheat (*Triticum aestivum* L.) Cultivars. World J Agric Sci 2008; 4(3): 351-8.

[62] Wakeel A, Steffens D, Schubert S. Potassium substitution by sodium in sugar beet (*Beta vulgaris*) nutrition on K☐fixing soils. J Plant Nutr Soil Sci 2010; 173(1): 127-34.
[http://dx.doi.org/10.1002/jpln.200900270]

[63] Zhu JK. Salt and drought stress signal transduction in plants. Annu Rev Plant Biol 2002; 53(1): 247-73.
[http://dx.doi.org/10.1146/annurev.arplant.53.091401.143329] [PMID: 12221975]

[64] Nassar ZM. Salt tolerance in fodder beet. Ph.D. Thesis, Fac. of Agric. 1989.

[65] El-Hawary MA. Effect of soil salinity and nitrogen fertilization on sugar beet. Ph.D. Thesis. Fac. of Agric.Al-Azhar Univ: Cairo. Egypt 1990.

[66] Eisa SS. Optimization of sugarbeet nutrition in sandy soil. Ph.D. Thesis, Fac. Agric.: Ain Shams Univ. Cairo, Egypt. 1999; pp. 50-60.

[67] Li J, Li H, Yang N, Jiang S, Ma C, Li H. Overexpression of monodehydroascorbate reductase gene from sugar beet M14 increased salt stress tolerance. Sugar Tech 2021; 23(1): 45-56.
[http://dx.doi.org/10.1007/s12355-020-00877-0]

[68] Dadkhah AR, Grrifiths H. The effect of salinity on growth, inorganic ions and dry matter partitioning in Sugar beet cultivars. J Agric Sci Technol 2006; 8: 199-210.

[69] Witkowski ETF, Lamont BB. Leaf specific mass confounds leaf density and thickness. Oecologia 1991; 88(4): 486-93.
[http://dx.doi.org/10.1007/BF00317710] [PMID: 28312617]

[70] Stevanato P, Gui G, Cacco G, *et al.* Morpho-physiological traits of sugar beet exposed to salt stress. Int Sugar J 2013; 115(1378): 756-65.

[71] Esechie HA, Al-Barhi B, Al-Gheity S, Al-Khanjari S. Root and shoot growth in salinity-stressed Alfaalfa in respone to nitrogen source. J Plant Nutr 2002; 25(11): 2559-69.
[http://dx.doi.org/10.1081/PLN-120014713]

[72] Hajiboland R, Joudmand A, Forouhi K. Mild salinity improves sugar beet (*Beta vulgaris* L.) quality. Acta Agric Scand B Soil Plant Sci 2007; 59(4): 229-305.
[http://dx.doi.org/10.1080/09064710802154714]

[73]   Hampe T, Marschner H. Effect of sodium on morphology, water relations, and net photosynthesis in sugar beet leaves. Z Pflanzenphysiol 1982; 108(2): 151-62.
[http://dx.doi.org/10.1016/S0044-328X(82)80066-4]

[74]   Khayamim S, Afshari RT, Sadeghian SY, Poustini K, Rouzbeh F, Abbasi Z. Seed Germination, plant establishment, and yield of sugar beet genotypes under salinity stress. J Agric Sci Technol 2014; 16: 779-90.

[75]   Zahra A. Evaluation of sugar beet monogerm O-type lines for salinity tolerance at vegetative stage. Afr J Biotechnol 2020; 19(9): 602-12.
[http://dx.doi.org/10.5897/AJB2019.17029]

[76]   Shaw B, Thomas TH, Cooke DT. Responses of sugar beet (*Beta vulgaris* L.) to drought and nutrient deficiency stress. Plant Growth Regul 2002; 37(1): 77-83.
[http://dx.doi.org/10.1023/A:1020381513976]

[77]   Bor M, Özdemir F, Türkan I. The effect of salt stress on lipid peroxidation and antioxidants in leaves of sugar beet *Beta vulgaris* L. and wild beet *Beta maritima* L. Plant Sci 2003; 164(1): 77-84.
[http://dx.doi.org/10.1016/S0168-9452(02)00338-2]

[78]   Scandalios JG. Oxygen stress and superoxide dismutase. Plant Physiology 1993; 101: 7-12.
[http://dx.doi.org/10.1104/pp.101.1.7]

[79]   Gzik A. Accumulation of proline and pattern of α-amino acids in sugar beet plants in response to osmotic, water and salt stress. Environ Exp Bot 1996; 36(1): 29-38.
[http://dx.doi.org/10.1016/0098-8472(95)00046-1]

CHAPTER 3

The Role of Functional Genomics to Fight the Abiotic Stresses for Better Crop Quality and Production

Neha Sharma[1,*], **Bharti Choudhary**[1] and **Nimisha Sharma**[2]

[1] *IILM University, Greater Noida, Uttar Pradesh 201306, India*

[2] *Indian Agricultural Research Institute, New Delhi, India*

Abstract: Plant quality, growth, yield and productivity are repeatedly affected by different abiotic stresses. It sometimes becomes a major upcoming threat to food security when the stress is on some staple crops. Stress-associated gene expression or no expression leads to abiotic stress tolerance, which is an outcome of complex signal transduction networks. Different plants have evolved with diverse, complex signaling networks concerning abiotic stresses. With the advancement of bioinformatics and functional genomics, in particular, many researchers have identified many genes related to abiotic stress tolerance in different crops, which are being used as a promising improvement in abiotic stresses. Different techniques of genome editing also play an important role in combating abiotic stresses. This chapter represents the knowledge regarding stress-tolerant mechanisms using technologies related to the field of functional genomics and may benefit the researchers in designing more efficient breeding programs and eventually for the farmers to acquire stress-tolerant and high-yielding crops to raise their income in the future.

Keywords: Abiotic stress, Crop quality, Crop production, Functional genomics.

1. INTRODUCTION

The extremes of temperature (high or low), drought, radiation, salinity, heavy metals, *etc.*, are collectively called abiotic stresses to the plants that strongly affect the plant growth, development, and eventually their quality and yield. Abiotic stress was found to be responsible for major crop losses throughout the world as they reduce yields and deteriorate the quality of the product by showing a negative impact on the actual potential of the crop. These stresses adversely affect crops in all means, in particular, physically and biochemically; also, the molecular expression of crops changes which ultimately affects the production of the crop

---

[*] **Corresponding author Neha Sharma** : IILM University, Greater Noida, Uttar Pradesh 201306, India; Email: nehabtc@gmail.com

Jen-Tsung Chen (Ed.)

[1 - 3]. It is anticipated that on account of environmental change, abiotic stresses might turn out to be more extreme and incessant. Dry spells and saltiness are turning out to be expanded in numerous places and may cause genuine salinization of over half of all arable terrains continuously by the year 2050. Subsequently, as a result of climbing temperatures and regular flooding for quite a long time, prolific horticultural land and harvest yields might diminish quickly. Moderate evaluations suggest that over 90% of the land in provincial regions is impacted by abiotic stress factors eventually during the developing season.

Details of various abiotic stresses and their effect on plants are well summarized in Table (**1**) [4 - 7].

Table 1. Details of abiotic stresses affecting the regular function of plant metabolism.

Stress	Mechanism	Major Effect on Plants
Drought	Water deficit and cellular dehydration (the decrease of the cell water content)	Reduced growth and metabolic activity with a reduction in yield
Heat	Enzymes inactivation and gradually denatured	Reproduction of plants, flowering, Male sterility, and abnormalities in the spikelets
Salt	Osmotic or water-deficit stress ionic toxicity (osmotic or water potential of the surrounding root zone is decreased by Na^+ and Cl^- solutes)	Inhibits plant growth and reduces the ability to uptake water and nutrients
Cold	Induce water leakage and lead to cell dehydration (inability to transport the water available from the soil to the living cell of leaf mesophyll)	Adversely affects plant growth and development, limits the geographical distribution of plant species

Ongoing innovative advances and the previously mentioned horticultural difficulties have prompted the rise of high-throughput devices to investigate and take advantage of plant genomes for crop improvement. The functional genomics-based approaches mean translating the whole genome, including genic and intergenic locales, which will thus give explicit procedures to have genetic improvement.

1.1. The Use of Functional Genomics in Studying Plant Physiology under Abiotic Stresses

1.1.1. Microarrays and MicroRNAs

The recent development in the field of computational biology with innovative omics techniques and technologies allows the identification of different genetic elements that contribute to various complex processes related to different stresses in crops [8]. In *Arabidopsis*, *AtPARP2* and *AREB1/ABF2,* responsible for the increased tolerance to drought have been studied *via* 24K and 22K Affymetrix microarrays, respectively [9, 10]. The technique of microarrays fills the bridging gap between functional genomics and sequencing data, which helps to understand the involvement of various genes in different regulatory networks and signal transduction pathways associated with tolerance or resistance against multiple abiotic crop stresses [8, 11]. The functional analysis through the method of microarrays reveals a number of signaling, metabolic and cellular pathways which in turn regulate the many biological and developmental processes, *viz.*, abiotic stresses. For example, 24K Affymetrix microarray revealed *AtMYB41*control the primary metabolism and negative regulation in *Arabidopsis* [12].

For miRNA analysis among crops, the development of different bioinformatic tools *viz.*, RNAhybrid, TarHunter, miRanda, MirGeneDB, and RNA22 [13], can detect miRNA targets and help in structure prediction. In *Arabidopsis*, miR398 was the first-ever reported miRNA to be involved in the regulation of auxin signaling which is related to stress tolerance. In the case of plants at various stages of development, miRNAs proved to be an important regulator of gene expression. Theoretically, Watson-Crick base pairing is hampered with miRNAs that lead to inhibition of translation either by cleavage of the target, which ultimately disrupts the function of genes at both transcriptional and post-transcriptional levels. For salinity stress tolerance, many active genes which are involved in transcription and other metabolic pathways were identified in wheat [14]. Hichri and co-workers [15] proved that the transcription factor SlWRKY3 is involved in salt stress tolerance in tomatoes and revealed that there is a significant relation between salinity stress and transcription factors from the family of the zinc finger. In *Oryza sativa*, miR156, miR158, miR159, miR397, miR398, miR482.2, miR530a, miR1445 on inducible genes, including *SalT* (LOC_Os01g24710) and *OsLEA3* (LOC_Os05g46480), were found to be responsible for drought tolerance, pathogen immune response and heat stress tolerance [16]. Wang and Long [17] found miR402 in *Zea mays* and found it to be responsible for seed germination and growth of seedlings under stress. In the case of wheat, it was observed that miRNAs (miR398) also play a substantial role in the down-regulation of gene expression when needed for adverse environmental stresses and oxidative activities [18]. Different inducible genes or their encoded proteins or transcription factors, including *Rd29A* (At5g52310) and CCAAT-binding transcription factors (*Arabidopsis*); CCAAT Binding Factor (CBF), Growth Regulating Factor (GRF), Cu/Zn superoxide dismutases (CSD1, CSD2) and TIR-NBS-LRR domain protein (*Medicago trunculata*), were found to be responsible for drought tolerance when interacting with different responsive miRNAs *viz*, miR164, miR169, miR389, miR393, miR396, miR397, miR402 (*Arabidopsis*) and miR169, miR396, miR398, miR2118 (*M. trunculata*).

1.1.2. Serial Analysis of Gene Expression (SAGE)

SAGE is an advanced transcriptomic tool used to produce a quantitation of thousands of messenger RNA populations that correspond to fragments of the transcripts. Cheng and co-workers [19] clearly stated that with the latest discoveries in the field of Next-Generation sequencing, SAGE had offered high throughput, reliable and responsive results to facilitate the simultaneous studies of thousands of transcripts. But the only problem we face with SAGE is that it works on short tags of 16 to 18 bases in length. Hence, to overcome this problem, based on longer reads, a newer method called SuperSAGE, allows the powerful and reliable expression of the genes [20]. Coming to SAGE, Cheng and co-workers [19] did a serial analysis of gene expression in *Coprinopsis cinerea* and revealed a transcriptomic switch during fruiting body development while studying abiotic stress tolerance. Therefore, with massive sequence analysis, High-throughput SuperSAGE or DeepSuperSAGE has given a platform to perform deep transcriptome analysis and multiplexing with reduced cost, effort and time. HT-superSAGE is very suitable for SOLiD sequencer and Illumina Genome Analyzer [20]. In many partially sequenced genomes, e.g. legumes SuperSAGE-Arrays were successfully used for high-throughput transcription profiling to procure differentially expressed genes in stress-susceptible and stress-tolerant genotypes. DeepSuperSAGE has been proven to be an amazing technique to detect the metabolic pathways for stress responsiveness in chickpeas [21]. SAGE has revitalized sequencing approaches of genomics and has given tremendous opportunities to molecular biologists for many emerging and unexploited analytical applications. Based on DeepSuperSAGE massive analysis of cDNA ends (MACE), stress-induced small RNome in legumes (pea and Medicago) has been elucidated in a new European FP7 project, for both biotic and abiotic stress to study epigenetics and changes due to DNA methylation [22]. Using 5' RACE (rapid amplification of cDNA ends) along with SAGE has helped to identify many new transcription initiation sites [23], which has proved to be more reliable, accurate, and high-throughput. SAGE has a good history of being employed in plants for the last many years, for example, in plant genomes, both biotic and abiotic stress responses, and the discovery of newly expressed regions [19].

1.1.3. RNA Sequencing

The highly sensitive technique of RNA-seq, where no prior gene information is required, has wider adaptability and utility for depth, and high throughput transcriptome analysis and is commendably efficient in exploring unknown transcripts. RNA seq technique provides endless opportunities for complete expression analysis *via* transcriptomic analysis of allele-specific regions, novel transcribed sequences, regions responsible for alternate splicing, editing of specific RNA, and strand-specific alterations [24]. The technique is cost-effective and can potentially explore the complete genome to design probe sets for the transcriptome analysis or identify non-coding RNAs [25]. An upregulation of almost 5545 genes for salinity tolerance and 4954 genes for drought tolerance was reported by Garg and co-workers [26] at various developmental stages of different chickpea varieties. Zhang and co-workers [27] revealed *via* RNA seq technique that during the expression of steroidal glycoalkaloid (SGA) in potatoes, both biotic and abiotic stress-responsive genes could be induced when introduced to light exposure. In rice during anthesis, using RNA seq, Gonza´lez-Schain and fellow researchers [28] proved the downregulation of many metabolic pathways and the upregulation of heat shock proteins in heat-sensitive varieties and heat-tolerant varieties, respectively. Differentially expressed genes for cold sensitivity and cold tolerance were found in sorghum genotypes using the RNA-seq technique [29]. Dugas and co-workers [30] found differential gene expression concerning different abiotic stresses and concluded a great association among different biochemical pathways, including plant defense.

1.1.4. RNAi

RNA silencing or RNA interference (RNAi) is a sequence-specific gene regulation technique leading to the post-transcriptional gene silencing (PTGS) elicited by the addition of double-stranded RNAs (dsRNAs), which inhibit specific gene expression. RNAi being one of the promising tools of functional genomics, leads to the knockdown of the expression of the mRNA by its degradation. The technique of RNAi was demonstrated to add abiotic stress tolerance traits in multiple crop species [31]. The technique of RNAi improved *M. sativa* by enhancing the target gene SPL13 (Squamosa Promoter Binding Protein-LIKE) and hence increased root length, increased stomatal conductance, added high levels of chlorophyll content and altered photosynthetic assimilation. It was clearly shown that the plants showed a reduction in water loss and enhancement in drought tolerance [32]. By using the RNAi technique, drought tolerance has been documented in *Sorghum bicolor*, *Solanum tuberosum*, *Triticum turgidum*, *Hordeum vulgare*, *Gossypium hirsutum*, *Oryza rufipogon* and *Cucumis sativus* and salinity tolerance identified in *Arabidopsis thaliana*, *Triticum aestivum*, *Gossypium hirsutum*, *Oryza sativa Raphanus sativus*, *Populus tomentosa*, *Zea mays* and *Cicer arietinum* [33]. The most recent technique of RNAi is spray-induced gene silencing (SIGS), in which the spraying of dsRNA on plant leaves, flowers, stems and fruits is done. This test could target pathogens to take up the sprayed dsRNA, and then they use their RNAi machinery to silence the targeted gene by amplifying, processing, and moving on to the silencing signal that ultimately leads to no expression. Based on RNAi techniques, several miRNAs were depicted to be involved in different cold stress and heat stress responses. In *Oryza sativa,* by targeting OsPCF5 and OsPCF8, miRNA tolerance against cold has been enhanced [34]. Xu and co-workers [35] deciphered chilling responsive miRNAs in *Glycine max* and Hivrale and co-workers [36] characterized *Panicum virgatum* for heat-responsive alterations in different miRNAs. *O. sativa* [37] and *Triticum aestivum* [38] have also proved heat-responsive alterations in different miRNA species. These researches prove the potential of RNAi technology to enhance plant development, growth and nutritional content. The possibility of an effect of several miRNA families, *viz.,* miR156, miR159, 169, and miR393, in response to abiotic stresses has been reported to regulate individually and/or together. With the advancement of genome sequencing and the availability of new genomes, scientists could better regulate these miRNA and their respective targets, which lead to crop improvement at different stages of the plant for different abiotic stresses. Hence, RNAi technology provides the most promising solutions to enhance crop quality, productivity, and tolerance to abiotic stresses. But, the only disadvantage of this approach is that sometimes the function of the gene is not completely inhibited, which leads to unintended off-target results [39].

1.1.5. CRISPR/Cas9

The CRISPR/Cas9 mediated genome editing technique is simple, specific and highly efficient and has a tremendous potential to develop abiotic stress-tolerant crop varieties for sustainable agriculture. In CRISPR/Cas9 genome editing, single guide RNA (sgRNA) guides the cleavage *via* Cas9, which in turn recognizes target DNA *via* Watson Crick base pairing and does the necessary editing. For the development of thermotolerant crops, CRISPR/Cas9 has proved to be the most successful approach [40, 41]. The CRISPR/Cas editing of the *slagamous-like 6* (*SlAGL6*) resulted in heat-tolerant parthenocarpic tomato fruits [42]. Cold stress-resistant rice was developed by editing ABA activated protein kinase 2 (SAPK2) gene [42]. The development of 53% efficient Ospin5b mutant for panicle length, 66% the gs3 mutant responsible for grain size, and 63% efficient Osmyb30 mutant for cold tolerance was credited to CRISPR/Cas9 in rice [43]. CRISPR/Cas depicted that C-repeat binding factor 1 (CBF1) protects the plant from chilling injury and aids to avoid electrolyte leakage [44]. Drought tolerance improved in wheat by CRISPR/Cas9 editing of dehydration-responsive element-binding protein 2 (TaDREB2) and ethylene-responsive factor 3 (TaERF3) [45]. The indica mega (MTU1010 dst mutant) rice cultivar generated by editing of drought and salt tolerance (*OsDST*) gene proved to produce broader leaves with decreased stomatal activity [46, 47] and hence the technique of CRISPR /Cas9 editing improved the rice salt and drought tolerance. Svitashev and co-workers [48] targeted *ALS1* and

ALS2 genes in maize by CRISPR/Cas to improve the mutant lines with herbicide resistance. CRISPR/Cas is a widely accepted and promising gene editing system for improving agricultural practices, which leads to next-generation precision breeding. CRISPR/Cas is a self-sufficient and independent gene editing system for the growing requirement of sustainability in agriculture to ensure global food security for upcoming generations.

1.1.6. Tilling and ECO Tilling

Targeting Induced Local Lesions IN Genomes (TILLING) technique provides an easy tool for the high-throughput analysis of a number of mutated organisms, and it goes in principle with almost all genes in which mutation is unavoidable [49]. Hexaploid wheat [50], sorghum [51], rice [52, 53], barley, maize [54] and soybean [55] are the few crop plants where tilling has been practically applied to demonstrate the stress response. For example, in leguminous species, few kinases were assessed *via* TILLING for salt stress response.

For polyploid species, EcoTILLING is considered to be the most utilized method to differentiate between homologous and paralogous genes. Recently, it was proved that EcoTILLING targets diversified stress responses, transcription factors, and different pathways leading to salt stress tolerance along with allelic variants to suggest the number of genes involved in salt stress [56, 57]. Many rice variants were exposed to drought stress to study the stress response and transcription factors responsible *via* EcoTILLING [58].

CONCLUSION

With the advancement of computational biology, functional genomics has been widely exploited, which has improved our understanding of the signaling pathways that work in stress tolerance among plants. Next-generation sequencing (NGS), genome-wide association (GWAS) studies, high-throughput proteomics, gene-editing mechanisms, gene silencing systems, and many bioinformatic discoveries have increased our interest in the use of functional genomics for combatting abiotic stress. At DNA Level, genetic interaction mapping, DNA/Protein Interaction, and DNA accessibility assays; At RNA Level, Microarrays, SAGE, RNA sequencing, and Massively Parallel Reporter Assays (MPRAs); At Protein level, Yeast two-hybrid system and AP/MS; At Mutagenesis and phenotyping level, Knockouts (Gene deletions), Site-directed Mutagenesis, RNAi and CRISPR/Cas could be used to get the pathways associated with the stress response. At all these different levels, plants were forced to adjust their transcriptome profile or change in gene expression, which led to establishing a new gene network and hence a change in gene function. Stress-related proteins are the main requisite players for stress signaling to combat abiotic stress, along with different miRNAs, transcription factors, metabolites, and hormones as they regulate gene expressions. This chapter will provide in-depth knowledge to researchers and scientists about the genes participating in regulatory networks and signal transduction associated with tolerance or resistance and recent advances in functional genomics for the betterment of crop production and quality against multiple abiotic stresses.

REFERENCES

[1]   He M, He CQ, Ding NZ. Abiotic stresses: general defences of land plants and chances for engineering multi stress tolerance. Front Plant Sci 2018; 9-1771.
         [PMID: 10.3389/fpls.2018.01771]

[2]   Sharma A, Soares C, Sousa B, *et al.* Nitric oxide-mediated regulation of oxidative stress in plants under metal stress: a review on molecular and biochemical aspects. Physiol Plant 2020; 168(2): 318-44.
         [PMID: 31240720]

[3]   Zafar SA, Zaidi SSA, Gaba Y, *et al.* Engineering abiotic stress tolerance *via* CRISPR/ Cas-mediated genome editing. J Exp Bot 2020; 71(2): 470-9.
         [http://dx.doi.org/10.1093/jxb/erz476] [PMID: 31644801]

[4]   Xiong L, Schumaker KS, Zhu JK. Cell signaling during cold, drought, and salt stress. Plant Cell 2002; 14(Suppl) (Suppl. 1): S165-83.
         [http://dx.doi.org/10.1105/tpc.000596] [PMID: 12045276]

[5]   Daryanto S, Wang L, Jacinthe PA. Global synthesis of drought effects on maize and wheat production. PLoS One 2016; 11(5): e0156362.
         [http://dx.doi.org/10.1371/journal.pone.0156362] [PMID: 27223810]

[6]   Farooq M, Gogoi N, Barthakur S, *et al.* Drought stress in grain legumes during reproduction and grain filling. J Agron Crop Sci 2017; 81-102.
         [http://dx.doi.org/10.1111/jac.12169]

[7]   Mathivanan S. Abiotic stress-induced molecular and physiological changes and adaptive mechanisms in plants. Abiotic Stress in Plants 2021; p. 315.
         [http://dx.doi.org/10.5772/intechopen.93367]

[8]   Raza A, Su W, Hussain MA, *et al.* Integrated analysis of metabolome and transcriptome reveals insights for cold tolerance in Rapeseed (*Brassica napus* L.). Front Plant Sci 2021; 12: 721681.
         [http://dx.doi.org/10.3389/fpls.2021.721681] [PMID: 34691103]

[9]   Gu Z, Pan W, Chen W, *et al.* New perspectives on the plant PARP family: *Arabidopsis* PARP3 is inactive, and PARP1 exhibits predominant poly (ADP-ribose) polymerase activity in response to DNA damage. BMC Plant Biol 2019; 19(1): 364.
         [http://dx.doi.org/10.1186/s12870-019-1958-9] [PMID: 31426748]

[10]   Rasheed S, Bashir K, Matsui A, Tanaka M, Seki M. Transcriptomic analysis of soil-grown *Arabidopsis thaliana* roots and shoots in response to a drought stress. Front Plant Sci 2016; 7: 180.
         [http://dx.doi.org/10.3389/fpls.2016.00180] [PMID: 26941754]

[11]   Ghorbani R, Alemzadeh A, Razi H. Microarray analysis of transcriptional responses to salt and drought stress in *Arabidopsis thaliana.* Heliyon 2019; 5(11): e02614.
         [http://dx.doi.org/10.1016/j.heliyon.2019.e02614] [PMID: 31844689]

[12]   Padmalatha KV, Dhandapani G, Kanakachari M, *et al.* Genome-wide transcriptomic analysis of cotton under drought stress reveal significant down-regulation of genes and pathways involved in fibre elongation and up-regulation of defense responsive genes. Plant Mol Biol 2012; 78(3): 223-46.

[http://dx.doi.org/10.1007/s11103-011-9857-y] [PMID: 22143977]

[13]   Pervaiz T, Amjid MW, El-kereamy A, Niu S-H, Wu HX. MicroRNA and cDNA-microarray as potential targets against abiotic stress response in plants: advances and prospects. Agronomy (Basel) 2021; 12(1): 11.
[http://dx.doi.org/10.3390/agronomy12010011]

[14]   Kawaura K, Mochida K, Ogihara Y. Genome-wide analysis for identification of salt-responsive genes in common wheat. Funct Integr Genomics 2008; 8(3): 277-86.
[http://dx.doi.org/10.1007/s10142-008-0076-9] [PMID: 18320247]

[15]   Hichri I, Muhovski Y, Žižková E, *et al.* The *Solanum lycopersicum* WRKY3 transcription factor SlWRKY3 is involved in salt stress tolerance in tomato. Front Plant Sci 2017; 8: 1343.
[http://dx.doi.org/10.3389/fpls.2017.01343] [PMID: 28824679]

[16]   Yu Y, Wu G, Yuan H, *et al.* Identification and characterization of miRNAs and targets in flax (*Linum usitatissimum*) under saline, alkaline, and saline-alkaline stresses. BMC Plant Biol 2016; 16(1): 124.
[http://dx.doi.org/10.1186/s12870-016-0808-2] [PMID: 27234464]

[17]   Wang Y, Long LH. Identification and isolation of the cold resistance-related miRNAs in *Pisum sativum* Linn. J Liaoning Norm Univ 2010; 2: 27.

[18]   Lu Q, Guo F, Xu Q, Cang J. LncRNA improves cold resistance of winter wheat by interacting with miR398. Funct Plant Biol 2020; 47(6): 544-57.
[http://dx.doi.org/10.1071/FP19267] [PMID: 32345432]

[19]   Cheng CK, Au CH, Wilke SK, *et al.* 5′-Serial Analysis of Gene Expression studies reveal a transcriptomic switch during fruiting body development in Coprinopsis cinerea. BMC Genomics 2013; 14(1): 195.
[http://dx.doi.org/10.1186/1471-2164-14-195] [PMID: 23514374]

[20]   Matsumura H, Yoshida K, Luo S, *et al.* High-throughput SuperSAGE for digital gene expression analysis of multiple samples using next generation sequencing. PLoS One 2010; 5(8): e12010.
[http://dx.doi.org/10.1371/journal.pone.0012010] [PMID: 20700453]

[21]   Molina C, Rotter B, Horres R, *et al.* SuperSAGE: the drought stress-responsive transcriptome of chickpea roots. BMC Genomics 2008; 9(1): 553.
[http://dx.doi.org/10.1186/1471-2164-9-553] [PMID: 19025623]

[22]   Poltronieri P, Santino A. Non-coding RNAs in intercellular and systemic signalling. Front Plant Sci 2012; 3: 141.
[http://dx.doi.org/10.3389/fpls.2012.00141] [PMID: 22783264]

[23]   Wei T, Deng K, Liu D, *et al.* Ectopic expression of DREB transcription factor, AtDREB1A, confers tolerance to drought in transgenic *Salvia miltiorrhiza.* Plant Cell Physiol 2016; 57(8): 1593-609.
[http://dx.doi.org/10.1093/pcp/pcw084] [PMID: 27485523]

[24]   Ozsolak F, Milos PM. RNA sequencing: advances, challenges and opportunities. Nat Rev Genet 2011; 12(2): 87-98.
[http://dx.doi.org/10.1038/nrg2934] [PMID: 21191423]

[25]   Klepikova AV, Kasianov AS, Gerasimov ES, Logacheva MD, Penin AA. A high resolution map of the *Arabidopsis thaliana* developmental transcriptome based on RNA-seq profiling. Plant J 2016; 88(6): 1058-70.
[http://dx.doi.org/10.1111/tpj.13312] [PMID: 27549386]

[26]   Garg R, Shankar R, Thakkar B, *et al.* Transcriptome analyses reveal genotype- and developmental stage-specific molecular responses to drought and salinity stresses in chickpea. Sci Rep 2016; 6(1): 19228.
[http://dx.doi.org/10.1038/srep19228] [PMID: 26759178]

[27]   Zhang W, Zuo C, Chen Z, Kang Y, Qin S. RNA sequencing reveals that both abiotic and biotic stress-responsive genes are induced during expression of steroidal glycoalkaloid in potato tuber subjected to

light exposure. Genes (Basel) 2019; 10(11): 920.
[http://dx.doi.org/10.3390/genes10110920] [PMID: 31718041]

[28]   González-Schain N, Dreni L, Lawas LMF, *et al.* Genome-wide transcriptome analysis during anthesis reveals new insights into the molecular basis of heat stress responses in tolerant and sensitive rice varieties. Plant Cell Physiol 2016; 57(1): 57-68.
[http://dx.doi.org/10.1093/pcp/pcv174] [PMID: 26561535]

[29]   Chopra R, Burow G, Hayes C, Emendack Y, Xin Z, Burke J. Transcriptome profiling and validation of gene based single nucleotide polymorphisms (SNPs) in sorghum genotypes with contrasting responses to cold stress. BMC Genomics 2015; 16(1): 1040.
[http://dx.doi.org/10.1186/s12864-015-2268-8] [PMID: 26645959]

[30]   Dugas DV, Monaco MK, Olson A, *et al.* Functional annotation of the transcriptome of *Sorghum bicolor* in response to osmotic stress and abscisic acid. BMC Genomics 2011; 12(1): 514.
[http://dx.doi.org/10.1186/1471-2164-12-514] [PMID: 22008187]

[31]   Jagtap UB, Gurav RG, Bapat VA. Role of RNA interference in plant improvement. Naturwissenschaften 2011; 98(6): 473-92.
[http://dx.doi.org/10.1007/s00114-011-0798-8] [PMID: 21503773]

[32]   Arshad M, Feyissa BA, Amyot L, Aung B, Hannoufa A. MicroRNA156 improves drought stress tolerance in alfalfa (*Medicago sativa*) by silencing SPL13. Plant Sci 2017; 258: 122-36.
[http://dx.doi.org/10.1016/j.plantsci.2017.01.018] [PMID: 28330556]

[33]   Khare T, Shriram V, Kumar V. RNAi technology: The role in development of abiotic stress-tolerant crops.Biochemical, physiological and molecular avenues for combating abiotic stress tolerance in plants. Elsevier 2018; pp. 117-33.
[http://dx.doi.org/10.1016/B978-0-12-813066-7.00008-5]

[34]   Yang C, Li D, Mao D, *et al.* Overexpression of microRNA319 impacts leaf morphogenesis and leads to enhanced cold tolerance in rice (*O ryza sativa* L.). Plant Cell Environ 2013; 36(12): 2207-18.
[http://dx.doi.org/10.1111/pce.12130] [PMID: 23651319]

[35]   Xu S, Liu N, Mao W, Hu Q, Wang G, Gong Y. Identification of chilling-responsive microRNAs and their targets in vegetable soybean (Glycine max L.). Sci Rep 2016; 6(1): 26619.
[http://dx.doi.org/10.1038/srep26619] [PMID: 27216963]

[36]   Hivrale V, Zheng Y, Puli COR, *et al.* Characterization of drought- and heat-responsive microRNAs in switchgrass. Plant Sci 2016; 242: 214-23.
[http://dx.doi.org/10.1016/j.plantsci.2015.07.018] [PMID: 26566839]

[37]   Mangrauthia SK, Bhogireddy S, Agarwal S, *et al.* Genome-wide changes in microRNA expression during short and prolonged heat stress and recovery in contrasting rice cultivars. J Exp Bot 2017; 68(9): 2399-412.
[http://dx.doi.org/10.1093/jxb/erx111] [PMID: 28407080]

[38]   Kumar RR, Pathak H, Sharma SK, *et al.* Novel and conserved heat-responsive microRNAs in wheat (*Triticum aestivum* L.). Funct Integr Genomics 2015; 15(3): 323-48.
[http://dx.doi.org/10.1007/s10142-014-0421-0] [PMID: 25480755]

[39]   Gaj T, Gersbach CA, Barbas CF III. ZFN, TALEN, and CRISPR/Cas-based methods for genome engineering. Trends Biotechnol 2013; 31(7): 397-405.
[http://dx.doi.org/10.1016/j.tibtech.2013.04.004] [PMID: 23664777]

[40]   Nguyen HC, Lin KH, Ho SL, Chiang CM, Yang CM. Enhancing the abiotic stress tolerance of plants: from chemical treatment to biotechnological approaches. Physiol Plant 2018; 164(4): 452-66.
[http://dx.doi.org/10.1111/ppl.12812] [PMID: 30054915]

[41]   Biswal AK, Mangrauthia SK, Reddy MR, Yugandhar P. CRISPR mediated genome engineering to develop climate smart rice: Challenges and opportunities. Semin Cell Dev Biol 2019; 96: 100-6.
[http://dx.doi.org/10.1016/j.semcdb.2019.04.005] [PMID: 31055134]

[42]   Klap C, Yeshayahou E, Bolger AM, *et al.* Tomato facultative parthenocarpy results from Sl *AGAMOUS-LIKE 6* loss of function. Plant Biotechnol J 2017; 15(5): 634-47.
[http://dx.doi.org/10.1111/pbi.12662] [PMID: 27862876]

[43]   Zeng Y, Wen J, Zhao W, Wang Q, Huang W. Rational improvement of rice yield and cold tolerance by editing the three genes OsPIN5b, GS3, and OsMYB30 with the CRISPR-Cas9 system. Front Plant Sci 2020; 10: 1663.
[http://dx.doi.org/10.3389/fpls.2019.01663] [PMID: 31993066]

[44]   Li T, Yang X, Yu Y, *et al.* Domestication of wild tomato is accelerated by genome editing. Nat Biotechnol 2018; 36(12): 1160-3.
[http://dx.doi.org/10.1038/nbt.4273] [PMID: 30272676]

[45]   Kim D, Alptekin B, Budak H. CRISPR/Cas9 genome editing in wheat. Funct Integr Genomics 2018; 18(1): 31-41.
[http://dx.doi.org/10.1007/s10142-017-0572-x] [PMID: 28918562]

[46]   Ganie SA, Wani SH, Henry R, Hensel G. Improving rice salt tolerance by precision breeding in a new era. Curr Opin Plant Biol 2021; 60: 101996.
[http://dx.doi.org/10.1016/j.pbi.2020.101996] [PMID: 33444976]

[47]   Santosh Kumar VV, Verma RK, Yadav SK, *et al.* CRISPR-Cas9 mediated genome editing of *drought and salt tolerance* (*OsDST*) gene in *indica* mega rice cultivar MTU1010. Physiol Mol Biol Plants 2020; 26(6): 1099-110.
[http://dx.doi.org/10.1007/s12298-020-00819-w] [PMID: 32549675]

[48]   Svitashev S, Young JK, Schwartz C, Gao H, Falco SC, Cigan AM. Targeted mutagenesis, precise gene editing, and site-specific gene insertion in maize using Cas9 and guide RNA. Plant Physiol 2015; 169(2): 931-45.
[http://dx.doi.org/10.1104/pp.15.00793] [PMID: 26269544]

[49]   McCallum CM, Comai L, Greene EA, Henikoff S. Targeting induced local lesions IN genomes (TILLING) for plant functional genomics. Plant Physiol 2000; 123(2): 439-42.
[http://dx.doi.org/10.1104/pp.123.2.439] [PMID: 10859174]

[50]   Chen L, Huang L, Min D, *et al.* Development and characterization of a new TILLING population of common bread wheat (*Triticum aestivum* L.). PLoS One 2012; 7(7): e41570.
[http://dx.doi.org/10.1371/journal.pone.0041570] [PMID: 22844501]

[51]   Xin Z, Li Wang M, Barkley NA, *et al.* Applying genotyping (TILLING) and phenotyping analyses to elucidate gene function in a chemically induced sorghum mutant population. BMC Plant Biol 2008; 8(1): 103.
[http://dx.doi.org/10.1186/1471-2229-8-103] [PMID: 18854043]

[52]   Cooper JL, Henikoff S, Comai L, Till BJ. TILLING and ecotilling for rice. Methods Mol Biol 2013; 956: 39-56.
[http://dx.doi.org/10.1007/978-1-62703-194-3_4] [PMID: 23135843]

[53]   Negrão S, Cecília Almadanim M, Pires IS, *et al.* New allelic variants found in key rice salt-tolerance genes: an association study. Plant Biotechnol J 2013; 11(1): 87-100.
[http://dx.doi.org/10.1111/pbi.12010] [PMID: 23116435]

[54]   Till BJ, Reynolds SH, Weil C, *et al.* Discovery of induced point mutations in maize genes by TILLING. BMC Plant Biol 2004; 4(1): 12.
[http://dx.doi.org/10.1186/1471-2229-4-12] [PMID: 15282033]

[55]   Cooper JL, Till BJ, Laport RG, *et al.* TILLING to detect induced mutations in soybean. BMC Plant Biol 2008; 8(1): 9.
[http://dx.doi.org/10.1186/1471-2229-8-9] [PMID: 18218134]

[56]   Comai L, Young K, Till BJ, *et al.* Efficient discovery of DNA polymorphisms in natural populations by Ecotilling. Plant J 2004; 37(5): 778-86.

[http://dx.doi.org/10.1111/j.0960-7412.2003.01999.x] [PMID: 14871304]

[57]   McCallum CM, Comai L, Greene EA, Henikoff S. Targeted screening for induced mutations. Nat Biotechnol 2000; 18(4): 455-7.
[http://dx.doi.org/10.1038/74542] [PMID: 10748531]

[58]   Henikoff S, Till BJ, Comai L. TILLING. Traditional mutagenesis meets functional genomics. Plant Physiol 2004; 135(2): 630-6.
[http://dx.doi.org/10.1104/pp.104.041061] [PMID: 15155876]

CHAPTER 4

Genetic Enhancement for Salt Tolerance in Rice

G. Padmavathi[1,*], **R. K. Singh**[2], **M.N. Arun**[3], **B. Umakanth**[1], **G.S.V. Prasad**[1] and **K. Muralidharan**[4]

[1] *Plant Breeding Department, ICAR-Indian Institute of Rice Research, Hyderabad 500030, India*

[2] *International Centre for Biosaline Agriculture Dubai, United Arab Emirates*

[3] *Agronomy Department, ICAR-Indian Institute of Rice Research, Hyderabad 500030, India*

[4] *Pathology Department, ICAR-Indian Institute of Rice Research, Hyderabad 500030, India*

Abstract: Rice is the major and dominant cereal food crop in the world. Salinity stress is the second most abiotic stress next to drought, limiting rice yield. Approximately 953 Mha area of the world is affected by salinity. Genetic improvement of salt tolerance is an efficient approach to achieving yield gain in salt-affected areas. Although high-yielding salt-tolerant rice varieties are developed, it is difficult to generate tailor-made adapted varieties through traditional breeding. Hence various crop improvement approaches are followed, including marker-assisted selection and transgenic technology apart from classical breeding. Numerous QTLs were identified through the molecular marker approach, and specifically, *Saltol* QTL was introgressed into elite lines through marker-assisted back cross-breeding, and improved salt-tolerant varieties were bred. Genetic engineering tools are also amply employed whereby the genes underlying various biochemical/physiological processes such as ion and osmotic homeostasis, antioxidation, signaling, and transcription-associated with increased tolerance were characterized, validated, and used to develop salt-tolerant lines of rice. Yet, a clear relationship between expected gains in salt tolerance *in vitro* has often not been observed in the field in terms of grain yield. Hence, an integrated approach involving molecular breeding and conventional breeding would certainly pave the way to enhance salt tolerance in rice.

Keywords: Rice, Salinity, Alkalinity, Screening techniques, Salt tolerance, Mechanisms of tolerance, Breeding, QTLs, Marker-assisted selection, Transgenics.

1. INTRODUCTION

Rice is the staple and major cereal food crop for the majority of the global population. It is cultivated in an extensive range of climatic conditions and is

* **Corresponding author G. Padmavathi:** Plant Breeding Department, ICAR-Indian Institute of Rice Research, Hyderabad 500030, India; E-mail: padmaguntupalli6@gmail.com

Jen-Tsung Chen (Ed.)

often exposed to multiple abiotic stresses, *viz.,* high temperature, cold, drought, high salinity and alkalinity, submergence, excess water, and mineral deficiency or toxicity. Salt stresses, including salinity and alkalinity stresses, are the most often encountered in problem soils, limiting rice production in irrigated and rainfed ecologies. It is estimated that salinity affects 950 m ha of arable land and more than 20% of irrigated land globally [1, 2]. Salinization of land and water severely affects agricultural productivity [3]. Breeding superior salt-tolerant varieties appear to be the most promising approach to cultivating rice, even in salt-affected lands. The stress-tolerant varieties are expected to provide a yield increase of about 2t ha^{-1} [4].

Although rice varieties with improved salt tolerance are generated through traditional breeding strategy, the genetic gain obtained is not encouraging due to the polygenic nature, involvement of complex mechanisms of salinity tolerance with multiple physiological and biochemical traits, high environmental influence, presence of low genetic variability for salt tolerance [5]. With the intervention of molecular marker analysis, it has become possible to analyze quantitative traits, including salt tolerance, and a number of QTLs were detected for salt tolerance in rice. The *Saltol* QTL derived from Pokkali [6, 7] controlling Na^+/K^+ homeostasis and the *SKC1* gene from Nona Bokra [8] are noteworthy, offering increased seedling stage salinity tolerance. The *Saltol* QTL is introgressed into local popular rice varieties for the accelerated development of a few new salt-tolerant varieties [9 - 13] through marker-assisted back cross-breeding (MABB) with improved salt tolerance. However, the MABB approach could not impart significant expected yield gain under prolonged salt stress during reproductive stage tolerance.

The transgenic approach involves the manipulation of genes that encode the synthesis of compatible organic solutes, antioxidants, Na and K transport proteins, antioxidants, detoxifying genes, and late embryogenesis abundant proteins found to impart salt tolerance in crops [14]. But variety development through a transgenic approach for cultivation needs further investigation. The present review discusses the impact of salt stress in rice production, its effects on the morphological, physiological, biochemical, and genetic characteristics in rice, screening, and mechanisms contributing to salt tolerance, and reviews the progress made in breeding salt tolerance coupled with molecular breeding and genetic engineering strategies to cope up with salt stress.

1.1. Classification of Problem Soils

Broadly the salt-affected problem soils can be classified into two types, *viz.*, saline and alkaline. Carbonates and bicarbonates of sodium and magnesium are the dominating anions in alkaline soils, while chlorides and sulfates of sodium and magnesium are often seen in saline soils. Soil salinity is characterized by soluble salts of high concentration and measured in terms of electrical conductivity (ECe). The soil exhibiting >4 dS m^{-1} ECe, <8.5 pH, and <15% exchangeable sodium percentage (ESP) is considered saline soil. The alkaline soils indicate >8.5 pH and < 4 dS m^{-1} ECe and >15% ESP. Saline-sodic soils record > 4 dS m^{-1} ECe and >15 percent ESP with varying pH having both saline and sodic properties [15, 16].

1.2. Halophytes *vs* Glycophytes

1.2.1. Sodium Extrusion

Broadly plants are categorized into halophytes (salt-loving) and glycophytes (salt-sensitive) depending on their ability to withstand high salt levels. Halophytes thrive well under high salt-affected areas, wherein Na Cl content is found to be as high as ~ 500 mM. Glycophytes include vegetable and grain crops that are sensitive to high salinity. Halophytes have certain inherent mechanisms to resist high salt concentrations. Halophytes 1) minimize salt entry into leaves by a preferential accumulation of K$^+$ and sodium export by salt glands, 2) compartmentalize excessive sodium in old leaves, leaf sheath, and stem as well as extrude salt by plasma membrane or vacuolar Na$^+$/H$^+$ antiporters, thus maintaining an increased cytosolic K$^+$/Na$^+$, 3) accumulate osmolytes such as glycine, betaine or proline for osmotic adjustment, 4) maintain high metabolism under detrimental salt concentrations, and 5) produce antioxidants to resist the oxidative damage. Glycophytes also show more or less similar strategies of salt tolerance adapted by halophytes but are not so well developed.

A few well-known halophytes are S*alicornia, Avicennia marina, Sesuvium portulacastrum, Suaeda salsa, Mesembryanthemum crystallinum, Thellungiella halophila* and *Porteresia coarctata*. Among them, *M.crystallinum* and *T. halophila* possess a formidable capacity to survive and reproduce at 500 mM NaCl [17] and extreme cold (-15° C). *P. coarctata*, a tetraploid (2n = 4x = 48), is regarded as a model plant among the wild relatives of rice. It is a potential source that can tolerate extended periods of saline water (20 to 40 dS m^{-1}) inundation. It can maintain a low Na$^+$/K$^+$ ratio within shoots and pump out excessive sodium through special glands on leaves, and sequester sodium to older and non-functional old leaves. Generally, crops, including rice, are glycophytes that are sensitive to salts even at lower levels of salinity (ECe < 4 dS m^{-1}) and cannot tolerate prolonged exposure to salinity. In extreme conditions of salt stress, glycophytes show mortality.

1.3. Effects of Salinity or Alkalinity on Rice

Salt or, more specifically, sodium chloride, forms the toxic component of salinity damage in crops. Salt in soil water causes deleterious effects on plant growth for two reasons. First, osmotic stress or water deficit induced in plants in response to high external salinity hampers water absorption through roots, leading to slower growth. Dehydration of plants occurs under severe stress. Second ionic stress wherein sodium and chloride ions get into the transpiration stream, thereby slowly building up toxicity that eventually interferes with the enzyme activity to reduce photosynthesis and protein synthesis and injure cells. Ultimately crop yield is reduced when grown on saline soils. Among cereals, rice is the most salt-sensitive crop. However, it is the only cereal crop recommended for desalinization as it grows under waterlogged conditions and aids salt to leach down [18]. Large genetic variability was reported among rice germplasm for salinity tolerance [19] as well as sodicity tolerance [20] despite its high sensitivity.

Salinity affects rice at any stage of its life cycle. Rice is highly tolerant to salt stress during germination compared to other stages of crop growth. Interestingly, rice shows acute tolerance at vegetative growth but is sensitive during flowering and reverts to tolerance at maturity [21].

Soil salinity exerts adverse effects on different physiological processes responsible for plant growth reduction [19]. The impact of salinity on yield is more pronounced at the reproductive stage observed at panicle initiation. But, seedlings at the vegetative stage show a good recovery from stress [22]. Salinity delays the emergence of panicle, and flowering reduces spikelet fertility and causes a reduction in the filled spikelets at elevated levels of salinity [23]. Salt stress causes leaf senescence [19] and panicle sterility, thereby reducing seed set [24]. The increased salt load reduces the root length, root numbers per plant, and shoot length [25]. Many important physiological traits such as stomatal conductance, leaf enlargement, and photosynthesis could be directly affected by reducing leaf tissue's turgor potential induced by the loss of water due to salt stress. Overall growth inhibition occurs as a consequence of injured transpiring leaves. The chlorophyll content of rice leaves was reduced due to the triggered enzyme activity of chlorophyllase under salt stress [26].

In general, salt stress reduces grain yield by adversely affecting yield components, *i.e.,* tiller number, panicle length, spikelet number, sterility, panicle weight, and harvest index. In several studies, plant growth retardation was found to be directly proportional to increased levels of salt content, and especially beyond the threshold level, crop yield decline begins linearly with an increase in salinity. For example, in rice, the threshold ECe is 3 dSm^{-1} ($\sim$ 30 mM NaCl) [27]. There are mechanisms in rice to tolerate excessive salts, which can be classified into osmotic tolerance, ion exclusion and tissue tolerance.

It is well established that there is great variability within the rice, and it can withstand pH of 9.2–10.2 concerning sodicity and ECe of 4.0–10.0 dS m^{-1} [20]. Further, in rice, a soil pH of 8.8-9.2 is considered non-stress, while 9.3 to 9.7 is medium stress and >9.7 is high stress. Soils are categorized as slightly saline (ECe: 2- 4 dSm^{-1}), moderately saline (ECe: 4-8 dS m^{-1} and highly saline (ECe: 8- 16 dSm^{-1}).

1.4. Screening Methods for Salinity or Alkalinity

As of today, different screening methods are available to judge the performance of rice genotypes when exposed to salt stress. They are naturally occurring saline or sodic fields, artificially stressed saline or sodic microplots, saline or sodic solution culture/hydroponics and saline or sodic trays. Field screening is a rapid method to identify promising salt-tolerant rice genotypes on a large scale, provided soil heterogeneity has been sufficiently cared for. However, the spatial heterogeneity and other associated soil-related stresses in the field complicate the evaluation of genotypes. Adopting appropriate experimental designs, such as the augmented design with increased replications of test genotypes and certain check varieties, would enable uniform exposure to varying stress conditions prevailing in the field.

Microplot screening is the most reliable method as it efficiently maintains a uniform level of salinity or alkalinity throughout the microplot thereby checking soil heterogeneity. Usually, sodicity (pH = 9.0, 9.6, and 9.9) and salinity (ECe 5–6, 8–9 and 10-12 dS m^{-1}) levels are maintained in them. In salinity screening experiments, IR 28 (sensitive check) and Pokkali (tolerant check) are used as control varieties, whereas VSR 156 (sensitive check) and CSR 36 (tolerant check) for alkalinity evaluation. Another method available is the Yoshida nutrient solution culture employed for mass screening of seedlings at the vegetative stage. The individual seedlings are scored for salt or alkali injury when once the seedlings of sensitive checks are dead. Scoring is done based on seedling vigor, survival, and salt /or alkali injury on a 1-9 scale Table (**1**) as per the standard screening procedure [28]. The plants scored with alkali or salt injury scores of 1 and 3 are considered highly tolerant and tolerant, respectively. The scores of 5, 7 and 9 are considered moderately tolerant, sensitive and highly sensitive, respectively.

Table 1. Salinity and alkalinity scoring (IRRI, 2013).

Score	Reaction	Symptoms
1	Highly tolerant	Normal plant growth and tillering
3	Tolerant	Nearly normal growth, reduced tillering and few leaves discolored (alkali)/whitish and rolled (salt)
5	Moderately tolerant	Growth and tillering reduced; maximum leaves discolored (alkali)/rolled (salt); only a few elongating
7	Sensitive	Complete cessation of growth, maximum leaves dried, few plants dying
9	Highly sensitive	Virtually all plants are dead or dying

1.5. Physiological and Biochemical Mechanisms of Salt Tolerance

To cope with rising salt, plants use a variety of physiological and biochemical defense mechanisms. Rice exhibits several salt tolerance mechanisms, including salt exclusion, *i.e.,* exclusion of Na^+ ions entry, sodium compartmentation in roots and older leaves, maintenance of low Na^+ concentration, K^+ Homeostasis, high cytosolic Na^+/K^+ ratio, tissue tolerance, osmotic adjustment through the accumulation of inorganic ions, control of ROS through antioxidants, stomata regulation to reduce transpiration and salt uptake, high initial seedling vigor and early maturity.

Plants suffer from both osmotic stress and ion toxicity when exposed to salinity **stress** [29]. To avoid a rapid accumulation of toxic salts in the cytoplasm or cell wall, tolerant plants restrict or exclude the entry of toxic sodium ions through roots by sodium transporters and evade from dehydration of plants. When this fails, plant cells attempt to compartmentalize the toxic cytosolic Na^+ that has already accumulated in cells into vacuoles driven by a plasma membrane and tonoplast Na^+/H^+ antiporters, thereby protecting cytoplasm from ion toxicity (ion homeostasis) [30]. Excessive sodium disrupts potassium homeostasis in cells. Salt-sensitive plants exhibit wilting due to reduced water uptake, growth inhibition, and subsequent death.

The physiological defense shown by plants is further augmented by an increase in the concentrations of various osmoprotectants. Plants under osmotic stress accumulate certain biochemical metabolites [31] known as osmolytes or osmoprotectants, like sugars, sugar alcohols, complex sugars (and quaternary amino acid derivatives. They protect cells from dehydration injury by turgor maintenance in roots and shoots in response to water deficit caused under salt stress. These osmolytes may also alleviate the toxic effects of reactive oxygen species [32].

Proline is one of the most specific osmolytes that a plant accumulates in the cytosol under stress and restores when stress is relieved. Thereby it balances the osmotic pressure of ions in plants under drought or salinity stress conditions Salt tolerant seedlings of rice were shown to synthesize more proline under salt stress [33]. However, many reports have indicated a negative association between salt and proline level in crops [34]. The probable reason for this could be the involvement of another predominant mechanism of tolerance in crop varieties/species other than osmotic adjustment.

During seed formation and in vegetative tissues, several stresses like salt, drought, cold and heat induce the accumulation of late embryogenesis abundant (LEA) proteins [35]. They are glycine-rich hydrophilic proteins found in plants, algae, fungi, and bacteria. They impart dehydration tolerance and protect the cellular membranes, proteins, enzymes, and mRNAs during water deficit and salt stress. Membrane proteins and other Na^+ transporters in cell membranes regulate the osmotic pressure in plant cells. They protect the plant from the drying effect of salt by Na^+ excretion from the root or transporting excess sodium ions from the cell cytoplasm into the vacuole for storage, thereby regulating the osmotic pressure in plant cells such as *OsCDPK7* gene in rice and dehydration-responsive element binding protein (*Os DREB1)* in rice.

Reactive oxygen species (ROS) are formed in plants, such as superoxide, hydrogen peroxide and hydroxyl radicals, under salt stress. These ROS cause oxidative damage to lipids and proteins in cell membranes and nucleic acids [36]. Plants have a defense mechanism of overproducing antioxidants or detoxifying enzymes through overexpression of ascorbate, glutathione, α tocopherol and carotenoids (antioxidants), superoxide dismutase, ascorbate peroxidase, catalase and other non-enzymatic-antioxidants-to-scavenge-ROS efficiently. In the Pokkali cultivar of rice, a greater accumulation of H_2O_2 in the tonoplast of the leaf sheath was found under salinity stress [37]. Upregulation of genes encoding enzymes APX, Catalase, Cu-Zn superoxide dismutase and Glutathione reductase associated with scavenging ROS during salt stress was reported [38].

Plants under stress show certain signal processes like activation of transcription factors that bind to DNA-regulatory sequences of target genes and modulate gene transcription resulting in increased tolerance for stress. The members of the NAC family, heat shock proteins, dehydration-responsive element binding factors, stress-associated proteins, *etc.,* are some of the transcription factors [39]. Overexpression of the stress-inducible transcription factors will increase the tolerance to salinity. In rice, it was suggested that *SNAC2* isolated from IRA109, a *japonica* rice act as a transcription factor improving tolerance to salinity [40]. *OsMYB6* gene encodes a stress-responsive MYB transcription factor and, when overexpressed, escalates dual tolerance to drought and salinity in rice. The expression of higher proline content, catalase and superoxide dismutase governed stress tolerance in transgenics compared to wild type without negative effects on the growth of transgenic lines grown under salinity [41].

Plants require potassium because it helps to maintain cellular osmotic equilibrium. The activity of potassium-specific transporters in plants maintains cellular potassium. These are responsible for the transportation of potassium from the soil and accumulation in cytosol. Under salinity stress, plants take up low amounts of Na^+ and high quantities of K^+ into the cytosol to maintain these ions at a low Na^+ to K^+ ratio. High-affinity potassium transporters such as *HKT1* mediate K^+ influx into the cells even under low external concentrations of the micromolar range [42]. HKT transporters would be involved in Na^+ transport across the plasma membrane of rice crop root epidermal cells. HKT sodium transporters function in two ways. They either absorb sodium from the soil solution to minimize sodium accumulation in leaves when K^+ is limited, or they remove sodium from xylem sap and load Na^+ into phloem sap to reduce sodium accumulation in leaves when K^+ is limited.

1.6. Genetic Studies and Conventional Breeding Approaches for Salt Tolerance

A comprehensive understanding of the genetic analysis of salt tolerance traits, its association with other component traits and heritability are essential for its manipulation in the breeding program. Salt tolerance is reported to be an oligogenic and polygenic trait in rice [43]. Subsequent genetic studies indicated a huge environmental influence and low heritability of salt tolerance traits. The genetics of salt tolerance is complex and governed by polygenes present at different loci in the rice genome [44]. In addition to additive and dominant gene effects and overdominance reported for the frequently analyzed physiological trait, namely the Na^+/K^+ ratio, a high value of narrow-sense heritability was also recorded for the Na^+/K^+ ratio along with grain yield [45].

There is very little information available on the inheritance of alkalinity tolerance. Several diallele analysis studies suggested the presence of additive and dominant genetic effects of various genes contributing to salt tolerance [46]. The polygenic nature of alkalinity tolerance was found in three rice crosses [47]. Grain yield displayed a positive significant correlation with filled grains per plant, K content, and test seed weight [48], while a negative correlation was displayed with spikelet sterility% and SES scores in character association studies [49]. A significant and positive correlation of Na^+/K^+ with salt injury scores during the seedling stage by analyzing physiological component traits with alkali injury scores was reported [50].

Early breeding efforts focused on improving locally domesticated landraces by pure line selections. Traditional varieties and local landraces grow naturally in salt-affected locations. Notable among them are Pokkali, Nona Bokra, and Kalarata, which originated in coastal areas of India and were extensively utilized as donors in breeding programs to generate locally adapted and improved cultivars. Globally Pokkali is the most extensively utilized donor in salt tolerance breeding programs [51]. Pokkali maintains a lower shoot Na^+ / K^+ ratio and grows relatively better under salinity. Unfortunately, linkage drag contributes to poor yield and grain quality, increased tallness and maturity of donors were often brought into new varieties from landraces and could not be eliminated during the selection process.

Consequent to the systematic breeding efforts, considerable progress has been made in breeding salt-tolerant varieties for cultivation across countries including India, Bangladesh, Egypt, Vietnam, Indonesia, Philippines, Sri Lanka, Vietnam, Myanmar, Sierra Leone, Gambia, Guinea Bissau, Guinea, and Senegal, *etc.* Table (**2**) [52].

Table 2. High-yielding salt-tolerant rice varieties released across countries (Singh *et al.*, 2021).

Country	Varieties
India	Jhona 349, PVR 1, Type 100, Narendra 2, Vikas, Mohan, Panvel 1, Usar 1, Panvel1, Panvel 2, CSR 4, Usar 1, CSR 10, AU 2, CST 7-1, Vyttilla 2, Vyttila 3, Vyttila 4, Lunishree, Sagara, Vyttila 5, Narendra Usar Dhan 2, CSR 13, CSR 27, CSR 30, Narendra Usar Dhan 3, Panvel 3, TRY 1, TRY (R) 2, Dandi, CSR 23, Sumati, Jarava, Naina, Narendra Usar Sankar Dhan 2, Vyttila 6, Bhuthnath, CSR 22, Amal Mana, Narendra Usar Dhan 2008, GNR-2, DRR Dhan 39, Vytilla 8, Luna Sampada, Luna Suvarna, CSR 43, Luna Barial, jagjeevan, Luna Sankhi, Binadhan 10, Binadhan 8, Gosaba 5, Chinsurah Nona-I, CARI Dhan-5, CSR 46, CSR 49, CSR 52,CSR 56, CSR 60, GNR-5, Pandu Ranga, Lavanya, Jyotsna, Auranga, NICRA Dhan
Bangladesh	Bina Dhan8, Bina Dhan10, Bina Dhan53, BRRI Dhan23, BRRI dhan40, BRRI dhan41 BRRI Dhan40, BRRI Dhan41, BRRI Dhan47, BRRI Dhan53, BRRI Dhan54, BRRI Dhan55, BRRI Dhan61, BRRI Dhan65, BRRI Dhan69, BRRI Dhan78
Egypt	Giza 177, Giza 178, Sakha 104, Sakha 111
Vietnam	OM 997, OM 576, OM 2717, OM 2517, OM 3242, AS 996, IR 2151
Indonesia	Pobbeli, Inpari 34 Salin Agritan and Inpari 35 Salin Agritan

(Table 2) cont.....

Country	Varieties
Philippines	IRRI 112 (Hagonoy), IRRI 113 (Bicol), IRRI 124 (Sipocot), IRRI 125 (Matnog), IRRI 126 (Naga), IRRI 128, NSIC Rc182 (Salinas 1), NSIC Rc184 (Salinas 2), NSIC Rc186 (Salinas 3), NSIC Rc188 (Salinas 4), NSIC Rc190 (Salinas 5 NSIC Rc290 (Salinas 6), NSIC Rc292 (Salinas 7),NSIC Rc294 (Salinas 8), NSIC Rc296 (Salinas 9), NSIC Rc324 (Salinas 10), NSIC Rc326 (Salinas 11), NSIC Rc328 (Salinas 12), NSIC Rc330 (Salinas 13), NSIC Rc332 (Salinas 14), NSIC Rc334 (Salinas 15), NSIC Rc336 (Salinas 16), NSIC Rc338 (Salinas 17), NSIC Rc340 (Salinas 18), NSIC Rc390 (Salinas 19), NSIC Rc392 (Salinas 20), NSIC Rc462 (Salinas 21), NSIC Rc464 (Salinas 22), NSIC Rc466 (Salinas 23), NSIC Rc468 (Salinas 24), SIC Rc470 (Salinas 25)
Sri Lanka	IR 2151, AC 69-1
Myanmar	Shwewartun, Sangankhan 4, Pyi Myanmar Sein, Shew ASEAN (CSR36), Sangankhan Sinthwelatt, Saltol Sin thew Latt
Sierra Leone	ROK 5, WAR 1, B 38 D2, Rohyb 183-B-5-B-1, Rohyb 162- B-1, ROK 37, ROK 35, NERICA L-20
Gambia	ITA 222, WAR 1, WAR 77-3-2-2, ITA 212, ARICA 11
Guinea Bissau	WAR 1, WAR 77-3-2-2
Guinea	WAR 73-1-M-1, WAR 77-3-2-2
Senegal	WAR 1, WAR 77-3-2-2, WAR 81-2-1-3-2
China	Changbai No. 6, Changbai No. 7, Changbai No. 9, Changbai No. 13, Jigeng No. 84, *and* Jinyuan 101
Japan	Mantaro rice, Kanto 51, Hama Minoru, Chikushiqing, *and* Lansheng
South Korea	Dongjinbyeo, Ganchukbyeo, Gyehwabyeo, Ilpumbyeo, Seomjimbyeo *and* Nonganbyeo

However, conventional breeding programs were impaired due to the genetic complexity of salt tolerance trait controlled by poly genes with smaller effects and a high degree of G X E, epistatic interaction with other genes, low heritability and complexity of tolerance mechanisms besides the involvement of several breeding cycles and selection. To achieve enhanced tolerance, breeding programs were advocated to incorporate multiple physiological traits associated with salt tolerance [53].

1.7. Doubled Haploid Approach

For a genetically complex trait such as salt tolerance, it is always advantageous to have highly homozygous doubled haploids (DH), which can be recovered quickly over conventional breeding. The successful production of the salt-tolerant lines obtained *via* anther culture-derived doubled haploid method was indicated [54]. Consistent efforts made by ICAR-IRRI collaborative network in India to develop promising rice varieties such as IR 51500-AC-17 (CSR 21), IR 511485-AC-1, and AC 6534-4 for salinity, AC 6533-3 for sodicity and AC 6534-1 (CSR 28) for salinity and sodicity stresses [55] through anther culture. CSR 28 has the credentials to be used as a salt-tolerant check for the global evaluation of the International Rice Soil Stress Tolerant Nursery (IRSSTN). Another variety IR 51500-AC11-1, was released as Bicol or PSBRc 50 for commercial cultivation in saline-prone areas of the Philippines [54].

1.8. Molecular Mapping of Salt Tolerance Genes/QTLs

Most of the agronomically important traits useful for rice improvement, including yield, resistance to pests and diseases, tolerance to abiotic stresses, and grain quality, are influenced by several polygenes. The genomic regions harboring these polygenes associated with quantitative traits are called quantitative trait loci (QTLs). The introduction of molecular markers has revolutionized rice research over the last three decades [56 - 59]. With these molecular developments, it became possible to develop DNA markers that can identify genomic regions and characterize and manipulate genetic variation for several agronomically important traits, particularly in QTLs of various crop species.

Several research groups have identified many QTLs of both major and minor, utilizing the mapping populations derived by crossing salt-sensitive and salt-tolerant parents. During the seedling stage, there was a significant increase in QTL mapping studies for salt tolerance in rice. QTL analysis provides information on their number, putative location in the genome, gene action (additive, dominance and interaction), phenotypic effects, the extent of variation, pleiotropic effects and epistatic interaction between the QTLs [60]. This knowledge on markers linked to QTLs is of immense value for selecting the desired individual plant having target genes or for introgression of the genes in crop improvement.

1.9. Seedling Stage QTLs

In rice, remarkable progress has been made in mapping QTLs for seedling stage tolerance in controlled conditions using various tolerance-related characteristics. To start with, a major QTL, namely *Saltol* maintaining a low Na^+/K^+ in shoot under salinity, was mapped on chromosome 1 for the first time by AFLP genotyping utilizing RILs derived from the cross Pokkali / IR29 [6] at IRRI, Philippines. This QTL accounted for 64-80% of the variability. One of the RILs designated as FL 478 (IR 66946- 3R-178-1-1), which showed early flowering and was more photosensitive than Pokkali, was identified to have high salt tolerance maintaining a lower Na^+/K^+ ratio than parents. On chromosome 1 *saltol* QTL was mapped [7] from the same RIL population of IR29 / Pokkali occupying 1.5 Mb flanked by RM140 and RM23. It was the most widely used improved donor in many salt tolerance breeding programs over due course of time.

Later *Saltol* region was fine-mapped using EST and SSR markers in BC_3F_4 generation of the same cross involving IR 29 and Pokkali, indicating Pokkali is the source of positive alleles in FL478 [61]. Different Pokkali alleles occupying a physical region between 11.0Mb and 12.2Mb region of *saltol* locus were suggested [62]. Further, it was reported that *saltol* may involve the *SKC1* gene of Nonabokra [8] located at 11.46 Mb.

Two significant QTLs for Na^+ and K^+ concentration of the shoots, *qSNC-7* and *qSKC-1,* were mapped on chromosomes 7 and 1, respectively in $F_{2.3}$ mapping population derived cross Nona Bokra and Koshihikari [63]. Further, q*SKC1* was cloned, which encodes an HKT transporter, *OsHKT8* [8]. *Saltol* locus was found to occupy a 4.5Mb region flanked by RM1287 and RM7075 on chromosome 1 between 10.8 and 15.3 Mb in the BC_3F_4 population of IR29/Pokkali [64].

Salt stress inhibits seed germination rate, initial vigor, length of shoot, and root and dry weight during the seed germination stage. Seedling growth and establishment ultimately determine biological yield and grain yield in crops [65]. In rice, salinity has a profound effect on reducing germination rate and seedling establishment [66]. Earlier reports indicated seven QTLs under salt stress in a DH population of IR64/Azucena for seed germination, seedling vigor, root length & dry weight of seedling, and 4 of them were located on chromosome 6 [67]. In Jiucaf iqing/ IR36 mapping population (F_2) three QTLs for salt tolerance rating (*qSTR1, qSTR5, qSTR9*), two QTLs for shoot dry weight (*qDWS 8, qDWS 9*) and two QTLs (*qNAK*) associated with salt tolerance rating, Na^+ /K^+ in roots and dry matter weight of shoots conferring 6.7 to 19.3% phenotypic variance were reported [68]. A set of 16 QTLs associated with seed germination ability was reported in a *japonica* variety Jiucaiqing [66].

A set of 16 QTLs was detected, conferring 23 physiological traits based on $F_{2.3}$ populations derived from Kalarata and Azucena [69]. The short arm of chromosome 1 was found to harbor colocalization of the majority of the QTLs conferring salinity tolerance and seedling vigor in the *Saltol* locus. Map-based cloning studies facilitated the isolation of a *QTL qSE3* from a Chinese landrace Jiucaiqing encoding a HAK transporter *OsHAK2,* and it fosters salinity tolerance during seed germination and seedling establishment [70]. Tightly linked QTLs for the shoot and root-related traits for salinity tolerance such as root length (*qRL2.1* and *qRL 2.2)*, shoot length (*qSL2*), shoot dry weight (*qSDW2*) and salt injury scores (*qSIS2*) occupying different regions on chromosome 2 were recorded in the donor Madina Koyo [71]. In a BIL population of Zhenshan 97/ACC9 cross, 23 QTLs during germination while 46 QTLs during the seedling stage were detected.

The major QTLs identified with >15% of phenotypic variance were *qGT 4, qGR-3d12* and *qGR1-3d12* during germination, while *qSH1, qSDW1* and *qSH 11* during the seedling stage [72].

In a BC_1F_2 population of the cross, WJZ/Nip for seed germination rate under salinity, a major QTL *qGR6.2* was fine mapped to a physical location of 65.9 kb wherein a candidate gene *LOC_OS06G 10650* encoding tyrosine phosphatase family protein was found [73]. The response of genotypes, when subjected to saline stress, is determined based on visual observations using SES scoring [28]. Two novel QTLs, *qST1* and *qST3*, were discovered on chromosomes 1 and 3, respectively, for salt injury scores [74] in a RIL population developed from the cross Milyang 23/ Gihobyeo.

The additional QTLs for major traits associated with salt tolerance in terms of SES scores were frequently detected on chromosome 1, occupying similar regions in an $F_{2,3}$ mapping population derived from a cross between Nona Bokra and Koshihikari. A set of 6 QTLs for salt injury in salinized nutrient solution were detected [75] in an $F_{2,3}$ mapping population of CSR 27 (tolerant) and MI-48 (sensitive). Three major QTLs governing salt injury were localized on chromosomes 1, 8, and 10 during the seedling stage [76]. A study employing F_5 RILs of At 354 x Bg352 identified 6 QTLs, designated as *qSSI*1 and *qSSI4* (salinity survival index), *qSL1* and *qSL4* (shoot length), *qSNK1* and *qSNK4* (shoot Na^+/K^+ ratio), on chromosomes 1 and 4, explaining large phenotypic variation ranging from 9 to 16% [77]. Using a RIL mapping population of IR 29/Pokkali three significant QTLs *qSIS1, qSIS12,* and *qSIS12* for salt injury scores were detected, accounting for 21.2% of the total variance in the population and Pokkali alleles at these QTLs contributed tolerance [78]. Novel QTLs were recorded in a RIL population derived from IR29/Hasawi, *i.e., qSES1.3, qSES1.4* for SES scores, *qSL1.2* and *qSL1.3* for shoot length, *qRL1* and *qRL1.2* for root length, *qFWsht1.2,* and *qDWsht1.2* for fresh and dry weights of shoots on chromosome1 particularly on long arm unlike *Saltol* and *SKC1* [79].

The comprehensive phenotypic performance, as revealed by SES scores of salinity, is, in turn, controlled by several physiological traits such as seedling survival, Na^+ uptake, K^+ uptake and plant growth rate. In other words, inhibiting Na^+ absorption, increasing K^+ accumulation in the cytosol and consequently lowering the Na^+/K^+ ratio is the primary method of tolerance adopted by plants in nature [80] under saline stress. Many genomic regions associated with low Na^+ uptake, high K^+ uptake and the optimal Na^+/K^+ ratio, the key salt tolerance determinants, have been reported in rice. As many as 16 QTLs were detected for the uptake of sodium and potassium using AFLP analysis in a RIL population and two NILs of IR 36 that differed in sodium transport [81]. Four of these QTLs were related to ion uptake in roots; one QTL for high Na^+ uptake (Q_{Na}) on chr #1, two QTLs for K^+ uptake (Q_{K1}, Q_{K2}) on chro# 6 and 9 and another QTL for Na^+/K^+ ratio ($Q_{Na\,k}$) on chr #4.

On the other hand, QTLs for uptake of Na^+, K^+ and Na^+/K^+ ratio were reported to share the same location on chromosome 1 in RILs derived from the cross IR 4630-22-2-5-1-3/IR 15324-117-3-2-2. However, the region was 2Mb away from the *Saltol* location [82]. Similarly, Lang and co-workers [83] reported QTLs associated with the uptake of K^+ located on different chromosomes, *i.e.,* 4, 6 and 9, suggesting the involvement of polygenes in an F_3 mapping population derived from DocPhung /IR 28. In another study, Lang and co-workers [84] detected QTLs on seven different chromosomes (1,2, 3,7, 9, 11 and 12) for Na^+ and K^+ uptake, Na^+/K^+ ratio, seedling survival and dry weight of root and shoot using F_8 RIL population of the cross between Tesnai 2 and CB genotypes. In another investigation, QTLs for physiological traits were found in the $F_{2:3}$ population of Tarommahali (tolerant) and Khazar (sensitive) under salinity stress. The phenotypic variation explained ranged from 9% for the Na^+/K^+ ratio to 38% for K^+ uptake [85].

A few salt-related QTLs were identified that expressed independently in seedling and tillering stages in the back cross-inbred line population (BC_2F_8) of IR64 (sensitive *indica*)/Binam (tolerant *japonica*) [86]. Using the BC_2F_5 population of a cross between Tarome-Molaei and Teqing, a significant QTL (*QKr1.2*) on chromosome 1 was detected for root K^+ content with a phenotypic variance of 30% conferring salt tolerance [87].

In a study on alkaline tolerance wherein, limited reports are available, the involvement of eleven significant QTLs associated with alkaline tolerance were mapped on four chromosomes # 2, 3, 7 and 11): two each for alkali injury scores, dry matter weight and panicle length; two for plant height, one each for sodium uptake, potassium uptake and panicle number. This study was conducted in $F_{2.3}$ mapping population derived from Samba mahsuri/CSR 27. The LOD scores ranged from 2.64 to 6.06, and the individual phenotypic contribution of QTLs (R^2%) ranged from 10.5 to 22% [88]. Similarly, in another experiment on alkaline tolerance, Liang and co-workers [89] mapped *qDSRs8-1* associated with dead seedling rate and *qDLRa5-3* associated with dead leaf rate under alkaline stress in a RIL population of the cross Yiai 1 / Lishuinuo. In another mapping population of 187 RILs of a cross between Bengal and Pokkali, eighty-five additive QTLs governing nine salinity-tolerant-related traits were reported [90].

In the proximity of the *qSKC1* region, QTLs for high shoot K^+ concentration was detected in Nona Bokra. Maintenance of adequate shoot Na^+/K^+ balance, Na^+

extrusion and compartmentation were advocated as probable methods governing tolerance to salinity in 'Nona Bokra' [91].

1.10. Reproductive Stage QTLs

The majority of the studies on rice salinity tolerance were conducted under controlled conditions during the seedling stage. Only a few research investigations were carried out to study reproductive stage salinity tolerance in field experiments compared to the seedling stage, which directly bears final grain yield. The seedling stage salinity tolerance reported in the experiments fails to correlate with the results of the reproductive stage under field conditions. A total of thirteen chromosomal regions with significant additive effects were detected on chromosomes 1,2, 3, and 8 in a population of CSR 27/MI 48. All QTLs had major effects of >10% of the phenotypic variation for Na^+ and K^+ concentration and seedling salt injury scores [92]. In another study on chromosome 8, colocalization of a QTL for spikelet fertility was found with *qClLV-8.1* for Cl^- content on chromosome 8 using the same population [93]. As many as five yield-related QTLs were recognized under salinity in donor 'Sadri' for different yield components [94]. Bimpong and co-workers [95] identified the two most stable QTLs from Hasawi donors for yield and their component trait, *i.e., qGY11* and *qTN11* across different genetic backgrounds of NERICA-L-19, Sahel 108 and BG90-2 on chromosome 11. In another investigation, 16 large-effect QTLs were reported during the reproductive stage for Na^+ content, pollen fertility, and Na/K ratio in flag leaf located on chromosomes 1, 7, 8, and 10 in donor Cheriviruppu [96].

In a RIL mapping population of CSR11/MI48, twenty-one QTLs located on rice chromosomes 1, 2, 3, 5, 6, 8, 9 and 12 were detected for grain yield under sodicity [97]. QTLs detected for traits conferring salinity tolerance often have epistatic effects and are affected by environments [98]. Seven significant and promising QTLs were detected (*qPLH $_{1.1}$, qTT $_{11.1}$, qTS $_{11.1}$, qNFS $_{2.1}$, qPFS $_2$.1* and *qYLD $_{2.1}$*) for the traits plant height, total tiller number, total spikelet number, filled spikelet number, percent filled spikelet and yield accounting for 13.4 to 18.4% phenotypic variation in an F_2 population derived from NSIC Rc222/BRRI Dhan 47 [99] which may probably serve in marker-assisted breeding.

A total of nine consistent and additive QTLs located on five different chromosomes (#1, 2, 3, 4 and 11) were identified across 2 years, *i.e.,* two for spikelet degeneration (*qDEG-S-2-2, qDEG-S-4-3*), five for spikelet sterility (*qSSI-STE-2-1, qSSI-STE-2-2, qSSI-STE-3-1, qSSI-STE-4-1, qSSI-STE-11-1*) one QTL each *qK-S-1-1* and *qSSI-Grain-2*-1 for K^+ content and stress susceptibility index for grain respectively accounting for 17–42% phenotypic variation [100] in a BC_3F_5 population of the cross Pokkali/IR 64. Functional genes encrypting a calmodulin-binding protein and potassium transporter documented within the QTL regions could be used in marker-assisted breeding programs. On chromosome 2, a grain yield QTL (*qGY-2*) was mapped along with QTL for the stress-tolerant index on chromosome 6, explaining 45% of phenotypic variance [101].

1.11. Wild Species - The Sources of Rice Salt Tolerance Genes/QTLs

Numerous studies on QTLs governing salt tolerance have been conducted in cultivated rice. Although the genus *Oryza* has 22 wild relatives, which constitute the potential genetic resources of novel genes with diverse adaptive salt tolerance mechanisms remain untapped. For example, the wild rice alleles at 13 QTLs on chromosomes 1, 2, 3, 6, 7, 9, and 10, explaining 87% phenotypic variation, were detected using introgression lines derived from *O. rufipogon* and cultivated rice Teqing with enhanced salt tolerance in the Teqing background [102]. Wang and his colleagues [103] identified two candidate genes, namely *LOC_Os05g31620* encoding calmodulin-related calcium sensor protein and *LOC_Os10g34730* encoding GRAM domain-containing protein at salt tolerance QTL regions (Chr # 1, 5, 7, 9-12). These two proteins showed increased expression of salt-tolerant lines within a set of 285 introgression lines derived from the cross *O.sativa* cultivar, 93-11 and *O.rufipogon.*

In another study employing backcross inbred lines developed from the cross 9311 and an African wild rice *O. longistaminata (*AA genome), a set of 18 *O.longistaminata* QTLs conferring salt tolerance was identified [104]. Among them, one QTL each for salt injury score (*qSIS2*), the water content of seedlings under salt treatment (*qWCSST2*) and the relative water content of seedlings (*qRWCS2*) were found to be colocalized on chromosome 2. Sequence and expression analysis revealed the candidate gene *MH02t0466900* encoding a cytochrome P450 86B1 responsible for increased salt tolerance. These wild rice QTLs can be incorporated into cultivated rice by crossing the respective tolerant BILs.

A consolidated view of salinity tolerance QTLs identified in various mapping populations in rice is presented Table (**3**). As many of them are observed to be minor with small effects, a few selective major QTLs for each trait are furnished in Table (**4**).

Table 3. QTLs identified using various mapping populations for salt tolerance in rice.

Mapping Population	Type of Mapping Population	QTLs Identified	Marker Type	Reference
Fixed-line (TN1)	Fixed-line (TN1)	1	RFLP	[105]
IR 29/ Pokkali	RILs (F_8)	11	AFLP, SSR	[6]
Tesanai 2/CB	RILs (F_8)	18	SSR	[84]
IR55178 (IR4630-22-2-5-1-3/IR15324-117-3-2-2)	RIL (F_6)	11	AFLP, RFLP and SSR	[82]
IR 29/ Pokkali	RIL (F_8)	3	RFLP, SSL and SSR	[7]
IR 29/ Pokkali	BC_3F_4	2	RFLP and SSR	[106]
Nona Bokra/Koshihikari	$F_{2:3}$	11	RFLP	[63]
Nona Bokra/Koshihikari	BC_3F_2	1	STS	[8]
Jiucaiqing/IR36	F_2	7	SSR	[68]
Milyang 23/Gihobyeo	RIL (F_{19})	2	EST and SSR	[74]
Nipponbare/Kasalath//Nipponbare	BILs (BC_1F_9 to BC_1F_{12})	2	RFLP	[107]
CSR 27/MI48	$F_{2:3}$	6	SSR	[75]
Tarommahalli/Khazar	$F_{2:3}$	32	SSR	[108]
IR64/Binam	BC_2F_8	35	SSR	[86]
TN1/CJ06	DH	14	SSR	[109]
Ilpumbyeo/Moroberekan	BC_3F_5	6	SSR	[110]
CSR27/ MI48	$F_2.F_3$	25	EST and SSR	[92]
Tarommahalli/Khazar	$F_{2:3}$	14	SSR	[85]
R29/Pokkali	RILs	27	SSR	[62]
CSR 27/MI48	RILs (F_7)	9	HvSSR and SSR	[93]
IR 29/ Pokkali	BC_3F_4	13	InDel, STS and SSR	[64]
Tarome-Molaei*3/ Tiqing	BC_2F_5	14	SSR	[87]
Teqing/Oryza rufipogon Accession	ILs	15	SSR	[102]
Shaheen Basmati,/Pokkali	$F_{2:3}$	22	SSR	[111]

Mapping Population	Type of Mapping Population	QTLs Identified	Marker Type	Reference
BRRI dhan40/IR61920-3B-22-2-1	F_2	3	SSR	[76]
IR26/Jiucaiqing	RILs ($F_{2:9}$)	7	SSR	[66]
IR 26/Jiucaiqing	RIL F $_{2:9}$	9	SSR	[112]
Gharib/Sepidroud	$F_{2:4}$	41	SSR and AFLP	[113]
Sadri/FL478	F_2	37	SSR	[94]
IR29×Hasawi	RIL (F_5)	7	SNP	[95]
Gharib/Sepidroud	F $_{2:4}$	17	AFLP and SSR	[114]
Dongnong425/Changbai10	BC_2F $_{2:3}$	13	SSR	[115]
IR64/IR4630-22-2-5-1-3	F_2	40	SSR	[116]
Pusa Basmati 1/Cheriviruppu	F_2	24	SSR	[96]
Ce258/IR75862-206-2-8-3-B-B-B ; Zhongguangxiang 1/IR75862-206-2-8-3-B-B-B	BC_1F_{10}	18	SSR	[117]
Bg352/At354	RIL (F_5)	83	SNP	[118]
Bengal/ Pokkali	RIL (F_6)	100	SNP	[90]
TCCP266/ Sakha102	F_2	10	SSR	[119]
93-11/O. rufipogon accession	BC_3	16	SSR	[103]
Bengal/ Pokkali	BC_4F_4	50	SSR and SNP	[120]
IR29/Hasawi	RIL ($F_{5:6}$)	34	SNP	[121]
IR29/Hasawi	RIL (F_5)	20	SNP	[122]
Jupiter/Nona Bokra	BC_3F_4	33	SSR	[123]
At354 x Bg352	F_5 RILs	6	SSR	[77]
Cheniere/ Nona Bokra	BC_3F_4	50	SSR	[91]
NSIC Rc222/BRRI dhan 47	F_2	3	SNP	[99]
Liang-You-Pei-Jiu (LYP9) parents 93-11 and PA64s	BC_5 F $_{2:3}$	8	SNP	[124]
Kongyu131/Xiaobaijingzi	RIL	2	SNP	[125]

Table 4. Significant QTLs with large effects mapped for salt tolerance in rice.

Trait	QTL	Chr#	Flanking markers	PV (%)	Reference
Saltol/K absorption	*Saltol*	1	P3/M9-8-*Saltol*	80.2	[6]
Saltol/Na absorption	*Saltol*	1	P3/M9-8-*Saltol*	64.6	
Saltol/Na -K Ratio	*Saltol*	1	P3/M9-8-*Saltol*	64.3	
K⁺ Uptake	*qKU-9*	9	E12M55-4 to RM205	19.6	[82]
Saltol Na⁺ Uptake	*saltol-qNU-1*	1	RM140-C1733S	39.2	[7]
Saltol K⁺ uptake	*saltol-qKU-1*	1	RM140-C1733S	43.9	
Saltol Na:K ratio	*saltol-qNKR-1*	1	RM140-C1733S	43.2	
Shoot Na⁺ concentration	*qSNC-7*	7	C1057-R2401	48.5	[63]
Shoot K⁺ concentration	*qSKC-1*	1	C1211-S2139	40.1	
Root K⁺ concentration	*qRKC-4*	4	C891-C513	21.6	
Seedling survival	*qSDS-1*	1	C813-C86	18	
Root Na⁺ concentration	*qRNC-9*	9	R1751-R2638	16.7	
Shoot length	*qSL-1.2*	1	R2414/C742	30	[126]
Shoot K⁺ concentration	*qSKC-1*	1	K159-K061	40.1	[8]
Na⁺/K⁺ ratio in roots	*qNAK-2*	2	RM318-RM262	19.3	[68]
Seedling tolerance	*qST1*	1	Estl-2 &RZ569A	27.76	[74]
Salt injury	*qSIS-3*	3	RM563-RM186	25.8	[75]
K⁺ uptake	*qKUP-8*	8	RM4955-RM152	38.22	[108]
Dry weight root	*qDWRO-9a*	9	RM1553-RM7424	27.43	
Dry weight shoot	*qDWSH-3*	3	RM1022-RM6283	23.21	
Na⁺ uptake	*qNAUP-1b*	1	RM8068-RM8231	22.17	
Na⁺ in stem at vegetative stage	*qNaSV-8.1*	8	RM563-RM186	53.63	[92]
Na⁺/K⁺ Ratio in stem at vegetative stage	*qNa/kSV-8.1*	8	RM145-RM5699	51.78	
SES tolerance score	*qSES9*	9	RM296-RM7175	55	[62]
Root K⁺ concentration	*qRKC2*	2	RM13197-RM6318	36	
Shoot Na-K Ratio	*qSNK1*	1	RM1287-RM10825	27	
SES score	*qSES1.2*	1	RM493	18.42	[64]
Potassium in root (K⁺ R)	*qKr1.2*	1	RM473A-RM128	30	[87]
Na⁺/K⁺ Ratio in root	*qNar/Kr5*	5	RM122-RM413	27.6	
Sodium in root (Na⁺R)	*qNas6*	6	RM3-RM528	24	
Potassium in Shoot (K⁺S)	*qKs5*	5	RM413-RM289	22	
Na⁺/K⁺ Ratio in shoot	*qNas/Ks1*	1	RM23-RM5	19	

(Table 4) cont.....

Trait	QTL	Chr#	Flanking markers	PV (%)	Reference
Relative root dry weight	*qRRW10*	10	RM271	26	[102]
Salinity tolerance	*SalTol8-1*	8	RM25-RM210	29	[76]
Germination percentage	*qGP-9*	9	RM219–RM7048	43.7	[66]
Imbibition rate	*qIR-9*	9	RM2144–RM3320	33.7	
Dry shoot weight	*qDSW6.2*	6	RM340-RM3509	23.9	[112]
Shoot K concentration	*qSKC-10b*	10	RM258-RM333	29.71	[113]
Standard tolerance ranking	*qSTR-9*	9	E36-M61-2-RM257	21.7	
Shoot dry weight	*qSDW-2*	2	RM279-RM6911	17.88	
Shoot Na/K ratio	*qSNK-8*	8	RM3572-RM404	15.12	
1000-grain weight	*qTGW8.1s*	8	RM80–RM281	30	[94]
Plant height	*qPH1.1s*	1	RM246–RM431	17	
Dry weight	*qDW2.1*	2	id2009319 - id2012453	42.3	[95]
Germination rate	*qGR-3-b*	3	E36-M61-11: E36-M61-11	19.23	[114]
Germination percentage	*qGP-4*	4	RM8213:E37-M60-3	15.65	
Shoot Na⁺ concentration	*qSNC-12*	12	RM28033 - RM1310	18	[115]
Root sodium concentration	*qRNC-9*	9	RM201 - RM215	16.6	
Plant height	*qPH4.1*	4	RM551-RM471	76.02	[116]
Tiller number	*qTN2.1*	2	RM279-RM424	30.84	
Root dry weight	*qRDW8.1*	8	RM210-RM458	26.48	
Plant height	*qPH1.1*	1	RM128-RM472	48.7	[96]
Panicle length	*qPL7.4*	7	RM180-RM3635	35.1	
Biomass	*qBM8.2*	8	RM3215-RM44	17.9	
Shoot dry weight	*qSDW10*	10	9898598-id10000153	45.3	[118]
Salinity survival index	*qSSI10*	10	9898598-id10000153	38.1	
Shoot Na⁺ concentration	*qSNC10*	10	9898598-id10000153	34	
Shoot Na⁺/K⁺ ratio	*qSNK 10*	10	9898598-id10000153	31.2	
Shoot K⁺ concentration	*qSKC10*	10	13069784-9922981	23.8	
Shoot-root ratio	*qSRR1.382*	1	S1-38286772 -S1-38611845	23.01	[90]
Salt injury score	*qSIS 7.17*	7	RM5793	16.989	[120]
Shoot dry weight	*qDWsht6.1*	6	id6016941-id6001397	48.4	[121]
Standard evaluation score	*qSES1.4*	1	id1023892-id1017885	42.3	
Shoot Na⁺/K⁺ ratio	*qSNK4*	4	RM518-RM5749	16	[77]
Salinity survival index	*qSSI4*	4	RM3843-RM280	15	
shoot dry weight	*qDWT8.1*	8	RM44-RM515	17.6	[91]

(Table 4) cont.....

Trait	QTL	Chr#	Flanking markers	PV (%)	Reference
Filled spikelet (%)	*qPFS2.1*	2	id2013434	18.4	[99]
Shoot length	*qSL6*	6	SNP6-99–SNP6-106	17.3	[124]
Shoot dry weight	*qSDW6*	6	SNP6-45–SNP6-78	16.3	
Root length	*qRL5*	5	SNP5-269–SNP5-289	23.7	

1.12. Marker-Assisted Introgression of *Saltol* in Rice

Despite the progress made in salt tolerance through traditional breeding methods, the longer breeding time and possible transfer of undesirable traits due to linkage drag causes yield loss and slow down breeding success. Consequent to the remarkable efforts made in the identification of a major *Saltol* QTL maintaining shoot Na^+ /K^+ homeostasis under salinity, its fine mapping and identification of tightly linked markers within the *Saltol* QTL region (AP3206, RM8094, and RM3412), flanking markers, *i.e.,* RM1287 and RM10694, RM493 and RM10793 markers [62], it was successfully transferred into wide popular varieties through MABB within the shortest time. Several researchers have successfully introduced the *Saltol* QTL into various rice varieties to breed new, improved salt-tolerant lines using BT7 [127], AS996 [128], Bacthom 7 [129], BRRI Dhan 49 [130], Novator [131], ADT43 [9], PB 1121 [132], Pusa Basmati 1 [10], Yukinko-mai [133], Pusa44 and Sarjoo 52 [11], Pusa Basmati 1509 [12] and Aiswarya [13] in India, Philippines, Bangladesh, Thailand, Vietnam and Senegal) Table (**5**).

Table 5. *Saltol* introgression from donor FL 478 (IR 66946-3R-178-1-1) into various elite rice varieties through MABB.

No.	Salt Tolerance QTL/ Gene	Recurrent Parent	Markers	Marker Type	Reference
1	*Saltol*	BT 7	RM493 RM3412b	SSR	[127]
2	*Saltol*	AS996	AP3206, RM3412 RM10793 RM10711	SSR	[128]
3	*Saltol*	Bacthom 7	RM3412 RM140 RM493	SSR	[129]
4	*Saltol*	BRRI dhan49	RM493	SSR	[130]
5	*Saltol*	Novator	RM493	SSR	[131]
6	*Saltol*	PB 1121	RM 3412	STMS	[132]
7	*Saltol*	ADT 43	RM3412 RM8094 *SKC 2a*	SSR	[9]
8	*Saltol*	Pusa Basmati 1	RM8094 RM493 RM10793	SSR	[10]
9	*hst1 gene in Saltol region*	Yukinko-mai	SNP	SNP	[133]

(Table 5) cont.....

No.	Salt Tolerance QTL/ Gene	Recurrent Parent	Markers	Marker Type	Reference
10	*Saltol*	Pusa Basmati 1509	AP3206, RM3412 RM10793	SSR	[12]
11	*Saltol*	Pusa44 and Sarjoo52	RM3412 AP 3206	SSR	[11]
12	*Saltol*	Aiswarya	AP 3206 SKC10 RM3412	SSR	[13]

The first and foremost investigation [127] exemplifies the introgression of *Saltol* of the tolerant donor, FL478 (IR 66946-3R-178-1-1) into a Vietnam sensitive variety, BT 7 through MABB with SSR markers RM3412b and RM493. Two of the BC_3F_1 lines possessing *Saltol* QTL upon field screening showed improved salt tolerance while retaining the agronomic performance of BT 2. Huyen and co-workers [128] employed MABB for successful introgression of *saltol* into Vietnam cultivar AS 996 variety using markers AP3206, RM3412, RM10793, RM10711 and generated five highly tolerant BC_2F_2 homozygotes for *Saltol*.

Through backcrossing, *Saltol* QTL was incorporated into an elite aromatic cultivar of Vietnam, namely Bacthom 7 [129]. For recovering the salt-tolerant genotypes, the inked markers RM140, RM3412 and RM493 were used till BC_3F_3 generations, while 81 polymorphic SSR markers to identify plants possessing 97% introgression of recurrent parent genome. In a similar effort, the salinity tolerance of Bangladesh variety BRRI dhan49 was improved through marker-assisted introgression of *saltol* employing RM 493 for foreground selection and a set of fifty-six polymorphic SSR markers for background selection [130]. The BC_2F_2 Introgression lines of BRRI dhan49 showed better salt tolerance upon screening in hydroponics. Incorporation of *saltol* QTL into a Russian variety Novator was carried out with the help of *saltol*-linked marker RM 493 [131]. Two lines, Nov-129 and Nov-148, were found to possess *Saltol* loci in the homozygous state with better salt tolerance. They could be explored as improved genetic sources in salt tolerance breeding programs.

A popular fine-grain Indian variety ADT 43 which is otherwise salt-sensitive, was incorporated with *Saltol* QTL utilizing RM 3412, RM 8094 and *SKC 2a* markers [9]. Many of the back cross inbred lines were not only positive for *saltol* but also exhibited yield gain over check variety and showed sodicity tolerance too. Through SNP marker-assisted selection, the *hst1* gene from the donor Kaijin was introgressed into a Japanese variety, Yukinko-mai (WT) [133] and generated a BC_3F_3 population named YNU31-2-4 with enhanced survival, increased shoot and root biomass and reduced Na^+ accumulation in shoots compared to the recurrent parent. The NILs generated by *Saltol* QTL introgression into Pusa44 and Sarjoo

52 rice varieties [11] exhibited lower salt injury scores as well as good ionic balance (Na^+/K^+).

The reports on incorporating *saltol* into basmati varieties for improving seedling stage salinity tolerance through MABB are scanty compared to non-basmati varieties. MAS has been successfully applied in developing salt-tolerant versions of Pusa Basmati 1, Pusa Basmati 1121 and Pusa Basmati 1509 varieties with the introgression of *saltol,* thereby stabilizing basmati production. Babu and co-workers [132] introduced *Saltol* into a Basmati rice cultivar, PB 1121. The STMS marker RM 3412, tightly linked to *Saltol,* was used for foreground selection. One of the BC_3F_3 derived NILs, Pusa1734-8-3-3, showed higher salinity tolerance in the seedling stage in the field retaining the yield and cooking quality of PB 1121. In another study, *Saltol* QTL was introduced into Pusa Basmati 1 through a marker-assisted backcrossing approach and salt-tolerant Pusa Basmati 1 NILs were recovered in the reproductive stage [10]. Salt tolerant NILs (BC_3F_4) of Pusa Basmati 1509 displayed yield increase along with good grain cooking quality under saline stress. *OsHKT1*;5 genes were found to regulate Na^+/K^+ homeostasis, thereby governing salinity tolerance with its higher expression levels under stress [12]. Considering the frequent occurrence of submergence along with salt stress in coastal areas, Nair and co-workers [13] transferred both *Saltol* QTL and *Sub1* gene in the Aiswarya variety leading to the development of elite advanced back cross-breeding lines with tolerance to salt as well as submergence up to 2 weeks.

1.13. Transgenic Approaches To Improve Salt Tolerance

The transgenic approach is a novel and quicker method for improving crop varieties. It facilitates the incorporation of desirable genes affecting qualitative and quantitative traits without any taxonomic barrier. Also, it allows the expression of an introduced gene or QTLs in a precisely controlled manner, both spatially and temporally. Initially, attempts have been made to insert novel genes known as transgenes identified from unrelated organisms, be it a bacterium or plant sources such as *E. coli,* halophytes and glycophytes. Later, with the rapid advances made in molecular biology, efforts shifted towards the overexpression of salt-responsive or related structural or functional genes involved in the process of tolerance in several crops [134]. They include manipulating genes encoding synthesis of compatible osmotica (proline, trehalose, glycine betaine, or polyamines, *etc.*), LEA proteins, sodium/potassium transport proteins, antioxidants, and detoxifying genes and transcription factors.

Overexpression of genes as a defense physiological mechanism is the main approach used in the genetic engineering of salt tolerance. One of them is glycine betaine which protects the photosystem II (PSII) complex in higher plants under salinity. The transgenic plants overexpressing GB synthesizing genes were found to be successful in the production of glycine betaine in plants in which GB was not normally produced and tolerated multiple stresses, including salinity stress [135].

Transgenic *japonica* rice cultivars of Zhonghua 8 and Zhonghua 10, upon transformation with *BADH* (betaine aldehyde dehydrogenase) cDNA of *Atriplex hortensis* were found to have a higher salt tolerance [136]. Transformed rice variety, *Nipponbare* with *CodA* gene for choline oxidase derived from the soil bacterium, *Arthrobacter globiformis* were found to have synthesized and accumulated glycine betaine in chloroplasts as well as in cytoplasm and boosted the tolerance to salt stress [137]. Increased tolerance in transgenic rice lines transformed with *E. coli mt1D* gene (mannitol-1 phosphate dehydrogenase) was shown due to the accumulation of mannitol [138].

A barley gene BADH was incorporated into rice *via Agrobacterium*-mediated genetic transformation and the transgenic rice lines exhibited a high salinity tolerance when compared to their counterpart wild-type plants [139]. An increase in K^+ content with the concomitant decrease in Na^+ and Cl^- content led to a gain in tolerance. Transgenic rice plants were developed using the *Agrobacterium*-mediated genetic transformation method wherein CMO (Choline monooxygenase) gene from spinach was incorporated, thereby inducing high salt and temperature tolerance [140]. Trehalose-6-phosphate synthase- phosphatase (*TPSP*) fusion gene could effectively increase tolerance to salt, drought, and cold in the transformed rice plants of Pusa basmati 1 [141]. The transformed Pusa Basmati-1 rice plants showed elevated levels of trehalose without yield penalty.

Genetically transformed rice plants were produced harboring a gene encoding a bifunctional fusion (TPSP) of the trehalose-6-phosphate (T-6-P synthase (TPS) and T-6-P phosphatase (TPP) of *Escherichia coli* introduced through *Agrobacterium*. Increased tolerance to salt, drought, and cold was found in transgenic rice due to the accumulation of trehalose in huge amounts [142]. Production of higher trehalose content in transgenics aided increased photosynthesis due to increased stomatal conductance and decreased photooxidative damage during stress from salt, drought and cold. It has been reported that transgenic expression of the trehalose-6-phosphate synthase gene (*OsTPS1I*) gene in transgenic plants confers salinity tolerance as seen by a rapid build-up of compatible solutes, seedling survival, and growth and reduced drying and photosynthetic maintenance efficiency [143]. Plant hormones such as Zeatin, gibberellic acid and indole-3-acetic acid were found in high concentrations promoting growth in *OsSUV3* transgenic rice than in the control plants in response to salt stress [144].

In transgenic plants, overexpression of *Oryza sativa* protein phosphatase 1a *(OsPP1a)* was associated with increased salt tolerance and upregulation of *SnRK1A, OsNAC5* and *OsNAC6* [145]. Overexpression of *HsCBL8 1* isolated from XZ166, a wild-barley in a rice variety, ZH 11 promoted salt tolerance in transgenic rice lines and was found to be associated with triggered proline content, reduced Na^+ uptake, and low Na^+/K^+ in shoot and root compared to non-transgenics [146].

LEA (Late embryogenesis abundant) genes are known to protect plants and play a role in abiotic stress damage repair pathways [147]. Overproduction of barley LEA gene, *HVA7,* when introduced into *Nipponbare* conferred enhanced drought and salt tolerance [148]. It was demonstrated that *HVA1* from Barley upon transfer into the Pusa basmati variety of rice using the *Agrobacterium*-mediated transformation system improved salt and drought tolerance on account of elevated levels of LEA3 accumulation in leaves [149].

Overexpression of *OsLEA3* in rice confers enhanced drought and salt tolerance in transgenics [150]. The introduction of the barley LEA protein gene, *HVA 1,* into rice caused a significantly enhanced tolerance to salinity and water deficit in the transgenic *japonica* rice variety Nipponbare [151, 152]. Similarly, wheat LEA genes, *PMA80* and *PMA1959,* were thought to be involved in desiccation tolerance in transgenic rice [153]. It was reported that a salt-tolerant transgenic *indica* rice variety Khitish was created by overexpressing the *Rab16A* gene, which is associated with the LEA protein. Transgenic plants had higher levels of *Rab16A* transcript and protein, suggesting increased salt tolerance by exhibiting low Na^+ content and K^+ loss, and enhanced proline content than wild-type plants [154]. The fact that the overexpression of *OsRab7* can increase salt tolerance was demonstrated in rice [155]. Through *Agrobacterium*-mediated transformation, the *OsRab7* gene was introduced into the japonica rice variety Zhonghua 11. The transformed rice plants showed overproduction of *OsRab7* protein and enhanced seedling growth and proline content under salt stress (200mM NaCl). More vesicles or vacuoles were observed in the root tip of *OsRab7* transgenic rice for the uptake of sodium.

As tolerance is likely to involve multiple genes, regulatory genes are efficient compared to structural genes in generating stress-tolerant plants. Multiple genes involved in stress response can be regulated at the same time by a single gene encoding a stress-inducible transcription factor [156], potentially improving tolerance to multiple abiotic stresses. Regulatory genes like calcium-dependent protein kinases (CDPKs), calcineurin B-like protein-interacting protein kinases (CIPKs), and transcription factors have been extensively utilized, which act upstream of salt-responsive genes. Any change in their expression would affect the downstream salt-related genes. The transgenic rice variety Notohikari over-expressed genes that modulate ion transport systems, such as that of Ca_2^+ dependent protein kinase (*OsCDPK7*) genes. The transgenic plants could survive in less than 200 mM NaCl compared to control plants that wilted for 3 days under stress [157]. Similarly, the *Agrobacterium*-mediated transformation of rice plants with *OsCDPK7* regulated by E12 promoter could correlate with the enhanced salt tolerance of transgenic plants [158].

Salinity-tolerant plants exhibiting sodium exclusion mechanism take up sodium through roots and pump the excessive sodium out through shoots to maintain cellular osmotic balance when exposed to salinity. Sodium includes, on the other hand, accumulates a high proportion of sodium in shoots obtained *via* a natural cation channel from the origin to the cell cytoplasm. Furthermore, certain tonoplast and plasma membrane Na^+/H^+ antiporters mediate energized sodium transport from the cytosol to the apoplast, within the vacuole, or back to the saline medium [159]. Plants can withstand salt stress by removing sodium ions from the cytosol *via* a plasma membrane or vacuolar Na^+/H^+ antiporters. The primary transport system for Na^+ compartmentalization into the vacuole or apoplast is the tonoplast antiporter, whereas the plasma membrane Na^+/H^+ antiporter primarily facilitates cellular Na^+ efflux.

The vast majority of transgenic studies have focused on genes involved in Na+ extrusion from the root or Na^+ compartmentalization. Certain vacuolar Na^+/H^+ antiporters that sequester Na^+ in vacuoles (*NHX1*) when overexpressed are known to have a role in improving tolerance to salinity in crops such as tomato and Brassica [160]. Overexpression of an *Arabidopsis thaliana* vacuolar Na^+/H^+ antiporter gene (*AtNHX1*) resulted in increased salt tolerance in transgenic *Arabidopsis* [161]. Transgenic rice overexpressing halophyte (*Atriplex gmelini*) gene (*AtNHX1*) and rice gene (*OsNHX1*) encoding Na^+/H^+ antiporter showed improved salt tolerance when compared to wild rice [162, 163]. The activity of the vacuolar Na^+/H^+ antiporter gene has neutralized the toxic effects of sodium ions by sequestering them into vacuoles.

SOD2, a plasma membrane Na^+/H^+ antiporter gene from yeast, was overexpressed in transgenic rice, resulting in increased salt tolerance [164]. The transformed seedlings displayed low Na^+ uptake, increased K^+/Na^+, high photosynthesis as well as decreased ROS generation. Improved tolerance by introducing three exogenous anti-apoptotic overexpressing genes, namely *AtBAG4, Hsp70,* and *p35,* in transgenic rice was demonstrated [165]. Photosynthesis, ion, and ROS maintenance system, crop growth rate, and yield components were all improved in the transformed plants following exposure to Na Cl stress. The salt-sensitive rice varieties transformed with the vacuolar Na^+/H^+ antiporter gene (*OsNHX1*) extracted from Pokkali, a land race that showed increased tolerance [166].

Transgenic rice overexpressing calcineurin B-like protein-interacting protein kinase, namely *OsCIPK15* in *japonica* rice Zhonghua, enhanced the protection against salt stress [167]. In another study, a serine-threonine protein kinase SAPK 4 acted as a regulatory factor during the acclimatization of plants to salt stress in rice. Transgenic rice plants overexpressing *SAPK 4* accumulated fewer Na^+ and Cl^- in responses to salt stress and performed better [168]. Similarly, when the transcription factor *OsDREB1* (rice DRE-binding protein 1) was over-expressed, it demonstrated improved tolerance to salt as well as drought and cold stress in rice [169, 170]. At the vegetative stage, transgenic rice overexpressing the drought tolerance transcription gene *SNAC1* demonstrated enhanced salt and drought tolerance [171]. Under high salinity conditions, the *SNAC2* gene was overexpressed in *japonica* rice Zhonghua 11, and the resulting transgenic rice exhibited enhanced germination and growth rates than the control [169]. Transgenic rice plants overexpressing the NAC transcription factor *ONAC045* showed improved drought and salt tolerance [170]. Overexpression of gene *SNAC1 TF* gene isolated from Pokkali, the traditional variety, was reported to impart both salinity and drought tolerance when incorporated into BRRI dhan55, a high-yielding variety in Bangladesh [172].

Overexpression of genes encoding antioxidants or detoxifying enzymes for scavenging active oxygen species was found to show enhanced salinity tolerance. When yeast mitochondrial *Mn-SOD* was overexpressed in rice, it increased salinity tolerance. This was indicated by the accumulation of ascorbate peroxidase and chloroplastic SOD in transgenic plants under stress conditions compared to wild type. This increased activity in the transformed rice was responsible for tolerance to salinity than the wild type [173]. Higher salinity stress tolerance was reported in transgenic lines of rice variety Zhonghua No.11 when Glutathione S-transferase and Catalase of *Suaeda salsa* were transferred through *Agrobacterium tumefaciens*-mediated transformation [174]. Nagamiya and Co-workers [175] showed that *Nipponbare* rice plants transformed with *katE* of *Escherichia coli* were found to be more tolerant under salinity stress. Prashanth and Co-workers [176] reported that overexpression of a cytosolic copper/zinc superoxide dismutase from *Avicennia marina in* Pusa Basmati 1 transgenic rice lines improved the tolerance of rice to high salinity and drought stresses.

Increased photorespiration caused by glutamine synthetase overexpression exhibited salt tolerance in transgenic rice plants [177]. To elude cell injury damage and nutrient depletion under salinity, plant cells maintain proper K^+ nutrition by potassium-specific transporters and a desirable Na^+/K^+ ratio in the cell cytoplasm [178]. Shaker family K$^+$ channel KAT1 *(OsKAT1)* of rice homolog was identified from a cDNA clone of *Nipponbare*. Seven expressed *OsHKT* transporters that could mediate Na$^+$ uptake in the rice cultivar *Nipponbare* were suggested [179]. In transgenic rice plants, overexpression of *OsKAT1* conferred salt stress tolerance [180].

For the first time, a gene *(PINO1)* for a novel salt MIPS (1-myo-Inosito--1-phosphate synthase) was reported from *Porteresia coarctata* [181]. The introduction of this gene into tobacco aided the transgenic plant in combating salinity-induced decrease in plant productivity and yield in transgenic tobacco plants [182]. It advocates a new approach to engineering salt tolerance in crops. *PDH45* gene overexpression in IR 64 transgenic rice was associated with the regulation of Na$^+$content, ROS production, Ca^{2+} cythomeostasis, cell viability, and cation transporters as well as improved salt tolerance [183]. Similarly, overexpression of transcription factor *OsMYB6* significantly contributed to salinity tolerance in *OsMYB6*-inserted transgenic plants [41]. Transgenic plants overexpressing MIPS coding genes *PcINO1* and *PcIMT1* gene (s) isolated from *Porteresia coarctata*, enabled improvement of saline tolerance in IR64 background [184]. It has been demonstrated that transgenic IR 64 rice plants containing the fusion gene from *E. coli* coding for trehalose-6-phosphate synthase/phosphatase were salt-tolerant without showing yield penalty by modulating the metabolism of amino acids and sugars [185]. The transgenic rice plants of the Vikas variety overexpressing *OsSOS1(NHX7)* through *Agrobacterium*-mediated transformation indicated enhanced salt tolerance exhibiting less membrane leakage, higher cell viability and higher K/Na ratio [186].

Recently, the overexpression of miR5505 in rice showed more substantial drought and salt tolerance than wild-type [187]. The ABA and stress responses involved small RINGH2 type E3 ligase *OsRF1* in *Oryza sativa,* revealing that the upregulation of *OsRF1* transcripts showed drought and salt tolerance and increment in endogenous ABA levels in rice [188]. Transgenic rice with the R2R3-MYB transcription factor *SiMYB19* overexpression increased salt tolerance during germination and seedling stages [189]. Despite the narratives on the development of transgenic rice lines for enhancing salinity tolerance Table (**6**), the progress is still in its infancy.

Table 6. Selected reports on saline-tolerant transgenics in rice.

Gene or Encoding Protein	Gene Source	Target Rice Variety	Reference	Tolerance
Osmolytes				
CodA (Choline oxidase)	*Arthrobacter Globiformis*	Nipponbare	[137]	Salinity and cold
	Arthrobacter pascens	TNG-67	[138]	Salinity
BADH	*Eschirichia coli*	-	[139]	Salinity tolerance
Ost A and Ost B	*Oryza sativa*	Pusa Basmati 1	[141]	Salinity and drought,
TPS and TPP (trehalose-6- phosphate phosphatase)	*E. coli*	Nakdong	[142]	Salinity, drought and cold
Choline monooxygenase (CMO)	Spinach (*Spinacia oleracea*)	Sasanishiki	[140]	Salinity and temperature
OsTPS1	*Oryza sativa*	ZH or Nipponbare	[143]	
OsSUV3	*Arabidopsis thaliana*	IR 64	[144]	Salinity
OsPP1	*Oryza sativa*	-	[145]	Salinity
PDH45	*Pisum sative*	IR 64	[183]	Salinity
Myo-inositol phosphate synthase (MIPS) coding genes*PcINO1* and *PcIMT1*	*Porteresia coarctata,*	IR 64	[184]	Salinity
Trehalose-6-phosphate synthase/phosphatase	*E. coli*	IR 64	[185]	Salinity and drought
Late embryogenesis abundant proteins (LEA) and Mitogen-activated protein kinase (MAPKs)				
OsCDPK7	*Oryza sativa*	Notohikari	[157]	Salinity, drought and cold
OsLEA3	*Oryza sativa*	-	[40]	Salinity
Serine-threonine protein kinase SAPK4	*Festuca rubra* ssp. Litoralis	IR 29	[168]	Salinity, drought and cold
HVA1	*Hordeum vulgare*	Nipponbare	[148]	Salinity and water deficit
HVA1	*Hordeum vulgare*	Pusa Basmati 1	[149]	Salinity and water deficit
Transcription factors				

(Table 6) cont.....

Gene or Encoding Protein	Gene Source	Target Rice Variety	Reference	Tolerance
Osmolytes				
DREB family genes	*OsDREB1A & OsDREB1B* from *Oryza sativa* and *DREB1A, DREB1B & DREB1C* from *Arabidopsis thaliana*	Kita-ake	[169]	Salinity, drought and cold
Calcineurin B-like Protein - interacting protein kinases (CIPKs like *OsCIPK03*, *OsCIPK12* and *OsCIPK15)*	*Oryza sativa*	Zhonghua 11	[167]	Salt, cold and drought
OsDREB1F	*Oryza sativa*	Nipponbare	[170]	Salt, drought, and low-temperature tolerance
OsNAC045	*Oryza sativa*	Guangluai4	[39]	Salinity and drought
Rab16A	*Oryza sativa*	Khitish	[154]	Salinity
OsRab7	*Oryza sativa*	Zhonghua 11	[155]	Salinity
AtBAG4, Hsp70 and *p35*	*Arabidopsis Citrus tristeza virus* Baculovirus	Nipponbare,	[165]	Salinity
SNAC1 TF	*Oryza sativa*	BRRIdhan 55	[172]	salinity and drought
HsCBL8	*Hordeum spontanum*	Zhonghua11	[146]	Salinity
OsMYB6	*Oryza sativa*	ZH 11	[41]	salinity and drought
OS RF1	*Oryza sativa*	Dongjin	[188]	Drought and salinity tolerance
MYB transcription factor SiMYB19 f	*Setaria italica*	Kitaake (Ki) rice	[189]	Salinity tolerance
miR5505	Dongxiang wild rice (*Oryza rufipogon*, DXWR)	Zhonghua 11	[187]	Drought and salt tolerance
Antioxidants				
Mn-SOD *(SOD)*	*Saccharomyces Cerevisiae*	Nakdong	[173]	Salinity
GST & CAT Glutathione-S Transferase and Catalase	*Suaeda salsa*	Zhonghua No. 11	[174]	Salinity and oxidative stress
Cu/Zn-SodI (SOD)	*Avicennia marina*	Pusa Basmati 1	[176]	Salinity, drought and oxidative stress

(Table 6) cont.....

Gene or Encoding Protein	Gene Source	Target Rice Variety	Reference	Tolerance
Osmolytes				
katE (Catalase)	*Eschirichia coli*	Nipponbare	[175]	Salinity
AtBAG4, Hsp70) and *p35*	*Arabidopsis, Citrus tristeza virus,* Baculovirus	Nipponbare	[165]	Salinity
Antiporters				
At NHX1 (Vacuolar Na^+/H^+ antiporter)	*Atriplex gmelini*	Kinohikari	[162]	Salinity
OsNHX1 (Vacuolar Na^+/H^+ antiporter)	*Oryza sativa*	Nipponbare	[163]	Salinity
SsNHX1 (Vacuolar Na^+/H^+ antiporter)	*Suaeda salsa*	Zhonghua 11	[174]	Salinity and water deficit
OsKAT1 (Shaker potassium channel)	*Oryza sativa*	Nipponbare	[180]	Salinity
OsNHX1 (Vacuolar Na^+/H^+ antiporter)	*Oryza sativa*	Binnatoa	[166]	Salinity
Na^+/H^+ antiporter (SOS1/NHX7)	*Oryza sativa*	Vikas	[186]	Salinity tolerance

CONCLUDING REMARKS

Salinity or alkalinity occurs due to intensive irrigation, high evaporation, and seawater intrusion. Saline/alkaline lands restrict the cultivation and impair the growth and productivity of rice. Improvement of salt tolerance coupled with high yield is the primary goal of salt tolerance breeding programs for rice globally. Although rice cultivars are moderately tolerant to salinity/alkalinity, wide traditional varieties, landraces and halophytic wild rice species are identified as potential donors. Advancement has been made in the elucidation of the genetic architecture of tolerance and incorporating saline/alkaline tolerance genes/QTLs into modern cultivars through classical breeding programs. The main obstacle in enhancing tolerance to salt stress has been the genetic and physiological complexity of the character, which is difficult to dissect and manipulate. The degree of tolerance has often been inadequate in the varieties developed, and the level of tolerance at one growth stage is genetically independent of tolerance at other stages. Hence, the reproductive stage is the key physiological stage thatdetermines the ultimate yield in rice plants. It should be given attention during selection in breeding programs.

The potential germplasm accessions possessing salt tolerance genes remain largely untapped. Research should focus on identifying and developing varieties that contain higher salt tolerance (ECe of >12 dS m^{-1}) and higher yield potential, and good grain quality. There is scope for domesticating a range of existing halophytes to improve salt tolerance. Molecular and biochemical tools are increasingly being used to mine genes from these distant species to overcome difficulties associated with crossability barriers.

It is clear that salt tolerance is a complex polygenic trait with low to medium heritability and strong GXE interaction and is not so easy to dissect and manipulate based on routine phenotypic selection. However, it has been demonstrated that a few QTLs, particularly *Saltol,* with significantly large effects, are employed in MAS.

The molecular marker-assisted breeding technique is a complementary tool to conventional breeding in the accelerated development of salt-tolerant rice genotypes combining two or more tolerance mechanisms. The current efforts of molecular assisted introgression of *Saltol* QTL into popular varieties, IR64, BR28, Swarna, PB 1121, BT 7, Yukinko-mai, *etc.*, across countries, show promising performance of derived lines under field stress reflecting the feasibility of this technique. However, certain limitations do exist in the implementation of QTLs in genetic improvement programs, such as low availability of major QTLs, inconsistency of QTL effects across locations, and genetic backgrounds effect, and undesirable epistatic effects during QTLs introgression. Proper care must be exercised to use appropriate mapping populations, preferably RILs, NILs, or DH, and validate major QTLs with large effects across locations and genetic backgrounds to select meaningful QTLs and identify key candidate genes for use in a MAS for salt tolerance.

The development of genetically engineered plants appears to be an alternate option to hasten salt tolerance breeding in plants. Many reports are available on the generation of transgenic rice with enhanced salt tolerance than the controls. Yet this method has not been a successful venture. In general, transgenic experiments are conducted under controlled conditions that are unlikely to occur in the field. In the transgenic approach, grain yield should be given due importance rather than banking upon component traits of salt tolerance.

Physiological investigations of salt tolerance suggest that a number of sub-traits determine the overall salt tolerance trait, and each one is, in turn, governed by the abundance of genes. These sub-traits are generally correlated with the minimization of the accumulation of harmful sodium and/or chloride ions, as well as potassium ion absorption from a high-salt environment. Hence it is necessary to exercise selection on physiological traits coupled with yield in breeding programs. The possibility of pyramiding genes underlying physiological traits that could act additively into elite rice varieties might accelerate the improvement of salt tolerance in rice.

REFERENCES

[1]   Solis CA, Yong MT, Vinarao R, *et al.* Back to the Wild: On a Quest for Donors Toward Salinity Tolerant Rice. Front Plant Sci 2020; 11: 323.
[http://dx.doi.org/10.3389/fpls.2020.00323] [PMID: 32265970]

[2]   Yang Y, Guo Y. Elucidating the molecular mechanisms mediating plant salt-stress responses. New Phytol 2018; 217(2): 523-39.
[http://dx.doi.org/10.1111/nph.14920] [PMID: 29205383]

[3]   Pareek A, Dhankher OP, Foyer CH. Mitigating the impact of climate change on plant productivity and ecosystem sustainability. J Exp Bot 2020; 71(2): 451-6.
[http://dx.doi.org/10.1093/jxb/erz518] [PMID: 31909813]

[4]   Ponnamperuma FN. Evaluation and improvement of lands for wetland rice production.In: Senadhira D, editor. Rice and problem soils in South and Southeast Asia .IRRI Discussion Paper Series No. 4. International RiceResearch Institute, Manila, Philippines; 1994.

[5]   Chinnusamy V, Jagendorf A, Zhu JK. Understanding and improving salt tolerance in plants. Crop Sci 2005; 45(2): 437-48.
[http://dx.doi.org/10.2135/cropsci2005.0437]

[6]   Gregorio GB. Tagging salinity tolerance genes in rice using amplified fragment length polymorphism (AFLP) 1997.

[7]   Bonilla P, Dvorak J, Mackell D, Deal K, Gregorio G. RFLP and SSLP mapping of salinity tolerance genes in chromosome 1 of rice (*Oryza sativa* L.) using recombinant inbred lines. Philippine Agricultural Scientist (Philippines) 2002; 85: 68-76.

[8]   Ren ZH, Gao JP, Li LG, *et al.* A rice quantitative trait locus for salt tolerance encodes a sodium transporter. Nat Genet 2005; 37(10): 1141-6.
[http://dx.doi.org/10.1038/ng1643] [PMID: 16155566]

[9]   Geetha S, Vasuki A, Selvam PJ, *et al.* Development of sodicity tolerant rice varieties through marker assisted backcross breeding. Electron J Plant Breed 2017; 8(4): 1013-21.
[http://dx.doi.org/10.5958/0975-928X.2017.00151.X]

[10]   Singh VK, Singh BD, Kumar A, *et al.* Marker-Assisted Introgression of Saltol QTL Enhances Seedling Stage Salt Tolerance in the Rice Variety Pusa Basmati 1. Int J of Genomics 2018; 12.
[http://dx.doi.org/10.1155/2018/8319879]

[11]   Krishnamurthy SL, Pundir P, Warraich AS, *et al.* Introgressed Saltol QTL Lines Improves the Salinity Tolerance in Rice at Seedling Stage. Front Plant Sci 2020; 11: 833.
[http://dx.doi.org/10.3389/fpls.2020.00833] [PMID: 32595689]

[12]   Yadav AK, Kumar A, Grover N, *et al.* Marker aided introgression of 'Saltol', a major QTL for seedling stage salinity tolerance into an elite Basmati rice variety 'Pusa Basmati 1509'. Sci Rep 2020; 10(1): 13877.
[http://dx.doi.org/10.1038/s41598-020-70664-0] [PMID: 32887905]

[13]   Nair MM, Shylaraj KS. Introgression of dual abiotic stress tolerance QTLs (*Saltol* QTL and *Sub1* gene) into Rice (*Oryza sativa* L.) variety Aiswarya through marker assisted backcross breeding. Physiol Mol Biol Plants 2021; 27(3): 497-514.
[http://dx.doi.org/10.1007/s12298-020-00893-0] [PMID: 33854279]

[14]　Ashraf M, Athar HR, Harris PJC, Kwon TR. Some prospective strategies for improving crop salt tolerance. Adv Agron 2008; 97: 45-110.
[http://dx.doi.org/10.1016/S0065-2113(07)00002-8]

[15]　Diagnosis and improvement of saline and alkali soils. US Department of Agriculture. Handbook 1954; (60): 160.

[16]　Eynard A, Lal R, Wiebe K. Crop Response in Salt-Affected Soils. J Sustain Agric 2005; 27(1): 5-50.
[http://dx.doi.org/10.1300/J064v27n01_03]

[17]　Inan G, Zhang Q, Li P, *et al.* Salt cress. A halophyte and cryophyte Arabidopsis relative model system and its applicability to molecular genetic analyses of growth and development of extremophiles. Plant Physiol 2004; 135(3): 1718-37.
[http://dx.doi.org/10.1104/pp.104.041723] [PMID: 15247369]

[18]　Lafitte HR, Ismail A, Bennett J. Proceedings of the 4s[th] International Crop Science Congress. Brisbane, Australia. 2004.

[19]　Munns R, James RA, Läuchli A. Approaches to increasing the salt tolerance of wheat and other cereals. J Exp Bot 2006; 57(5): 1025-43.
[http://dx.doi.org/10.1093/jxb/erj100] [PMID: 16510517]

[20]　Mishra B. Highlights of research on crops and varieties for salt affected so Technical Bulletin. Karnal, India: CSSRI 1996; p. 28.
[http://dx.doi.org/10.1016/S0378-4290(01)00128-9]

[21]　Shereen A, Mumtaz S, Raza S, Khan MA, Solangi S. Salinity effects seedling growth and yield components of different inbred rice lines. Pak J Bot 2005; 37(1): 131-9.

[22]　Asch F, Woperies MC. Responses of field-grown irrigated rice cultivars to varying levels of floodwater salinity in a semi-arid environment. Field crop Res 2001; 70(2): 12.

[23]　Zaibunnisa A, Khan MA, Flowers TJ. Causes of sterility in rice unsalinity stress.Prospects for saline agriculture. Netherlands: Kluwer Academic Publishers 2002; pp. 177-87.

[24]　Hasanuzzaman M, Fujita M, Islam MN, Ahamed KU, Nahar K. Performance of four irrigated rice varieties under different levels of salinity stress. Int J Integr Biol 2009; 6: 85-90.

[25]　Jiang XJ, Zhang SH, Miao LX, Tong T, Liu ZZ, Sui YY. Effect of salt stress on rice seedling characteristics, effect of salt stress on root system at seedling stage of rice. Beifang Shuidao 2010; 40: 21-4.

[26]　Abdullah ZKMA, Khan MA, Flowers TJ. Causes of sterility in set of rice under salinity stress. J Agronomy Crop Sci 2011; 87(1): 25-32.
[http://dx.doi.org/10.1046/j.1439-037X.2001.00500.x]

[27]　Research Databases 2008.http://www ars.usda.gov/Services/docs.html

[28]　IRRI Standard evaluation system for rice. 4th ed. Philippines: International Rice Research Institute 2013; p. 34.

[29]　Hasegawa PM, Bressan RA, Zhu JK, Bohnert HJ. Plant cellular and molecular responses to high salinity. Annu Rev Plant Physiol Plant Mol Biol 2000; 51(1): 463-99.
[http://dx.doi.org/10.1146/annurev.arplant.51.1.463] [PMID: 15012199]

[30]　Blumwald E, Aharon GS, Apse MP. Sodium transport in plant cells. Biochim Biophys Acta Biomembr 2000; 1465(1-2): 140-51.
[http://dx.doi.org/10.1016/S0005-2736(00)00135-8]

[31]　Chen TH, Murata N. Enhancement of tolerance of abiotic stress by metabolic engineering of betaines and other compatible solutes. Curr Opin Plant Biol 2002; 5(3): 250-7.
[http://dx.doi.org/10.1016/s1369-5266(02)00255-8]

[32]　Hong Z, Lakkineni K, Zhang Z, Verma DPS. Removal of feedback inhibition of delta(1)-pyrroline-

5-carboxylate synthetase results in increased proline accumulation and protection of plants from osmotic stress. Plant Physiol 2000; 122(4): 1129-36.
[http://dx.doi.org/10.1104/pp.122.4.1129] [PMID: 10759508]

[33] Chutipaijit S, Cha S, Sompornpailin K. Differential accumulations of proline and flavonoids in indica rice varieties against salinity. Pak J Bot 2009; 41(5): 2497-506.

[34] Handa S, Handa AK, Hasegawa PM, Bressan RA. Proline accumulation and the adaptation of cultured plant cells to water stress. Plant Physiol 1986; 80(4): 938-45.
[http://dx.doi.org/10.1104/pp.80.4.938] [PMID: 16664745]

[35] Sivamani E, Bahieldin A I, Wraith JM, *et al.* Improved biomass productivity and water use efficiency under water deficit conditions in transgenic wheat constitutively expressing the barley *HVA1* gene. Plant Sci 2000; 155(1): 1-9.
[http://dx.doi.org/10.1016/S0168-9452(99)00247-2] [PMID: 10773334]

[36] Van Breusegem F, Vranová E, Dat JF, Inzé D. The role of active oxygen species in plant signal transduction. Plant Sci 2001; 161(3): 405-14.
[http://dx.doi.org/10.1016/S0168-9452(01)00452-6] [PMID: 11166426]

[37] Wi SG, Chung BY, Kim JH, Lee KS, Kim JS. Deposition pattern of hydrogen peroxide in the leaf sheaths of rice under salt stress. Biol Plant 2006; 50(3): 469-72.
[http://dx.doi.org/10.1007/s10535-006-0073-6]

[38] Hong CY, Chao YY, Yang MY, Cheng SY, Cho SC, Kao CH. NaCl-induced expression of glutathione reductase in roots of rice (*Oryza sativa* L.) seedlings is mediated through hydrogen peroxide but not abscisic acid. Plant Soil 2009; 320(1-2): 103-15. a
[http://dx.doi.org/10.1007/s11104-008-9874-z]

[39] Zheng X, Chen B, Lu G, Han B. Overexpression of a NAC transcription factor enhances rice drought and salt tolerance. Biochem Biophys Res Commun 2009; 379(4): 985-9.
[http://dx.doi.org/10.1016/j.bbrc.2008.12.163] [PMID: 19135985]

[40] Hu H, You J, Fang Y, Zhu X, Qi Z, Xiong L. Characterization of transcription factor gene *SNAC2* conferring cold and salt tolerance in rice. Plant Mol Biol 2008; 67(1-2): 169-81. b
[http://dx.doi.org/10.1007/s11103-008-9309-5] [PMID: 18273684]

[41] Tang Y, Bao X, Zhi Y, *et al.* Li, Liu K. Overexpression of a MYB family gene, *OsMYB6*, increases drought and salinity stress tolerance in transgenic rice. Front Plant Sci 2019; 10: 168.
[http://dx.doi.org/10.3389/fpls.2019.00168] [PMID: 30833955]

[42] Sentenac H, Bonneaud N, Minet M, *et al.* Cloning and expression in yeast of a plant potassium ion transport system. Science 1992; 256(5057): 663-5.
[http://dx.doi.org/10.1126/science.1585180] [PMID: 1585180]

[43] Mishra B, Singh RK, Jetly V. Inheritance pattern of salinity tolerance in rice. J Genet Breed 1998; 52: 325-32.

[44] Flowers TJ. Improving crop salt tolerance. J Exp Bot 2004; 55(396): 307-19.
[http://dx.doi.org/10.1093/jxb/erh003] [PMID: 14718494]

[45] Gregorio GB, Senadhira D. Genetic analysis of salinity tolerance in rice (Oryza sativa L.). Theor Appl Genet 1993; 86(86): 333-8.
[http://dx.doi.org/10.1007/BF00222098] [PMID: 24193479]

[46] Saharay PK, Islam MA. Genetic analysis of salinity tolerance in rice. Bangladesh J Agric Res 2008; 33: 519-29.

[47] Singh RK, Mishra B, Jetly V. Segregations for alkalinity tolerance in three rice crosses. SABRAO J Breed Genet 2001; 33: 31-4.

[48] Mishra B. Breeding for salt tolerance in crops.Salinity management for sustainable agriculture: 25 years of research at CSSRI. Karnal, India: Central Soil Salinity Research Institute 1994; pp. 226-59.

[49]   Gupta D. Genetics of salt tolerance and ionic uptake in rice (*Oryza sativa L*). Ph.D Thesis. BRA University, Agra, India, 1999; p.112.

[50]   Padmavathi G, Surekha K, Ram Prasad AS. Correlation and path analysis of salt tolerance component traits in rice (*Oryza sativa L*). Poster presentation in 2nd International Rice Congress, New Delhi, India, 2006; 234, October 9-13.

[51]   Waziri A, Kumar P, Purty RS. Saltol QTL and Their Role in Salinity Tolerance. Austin J Biotechnol Bioeng 2016; 3(3): 1067.

[52]   Singh RK, Kota S, Flowers TJ. Salt tolerance in rice: seedling and reproductive stage QTL mapping come of age. Theor Appl Genet 2021; 134(11): 3495-533.
[http://dx.doi.org/10.1007/s00122-021-03890-3] [PMID: 34287681]

[53]   Sathish P, Gamborg OL, Nabors MW. Establishment of stable NaCl-resistant rice plant lines from anther culture: distribution pattern of K+/Na+ in callus and plant cells. Theor Appl Genet 1997; 95(8): 1203-9.
[http://dx.doi.org/10.1007/s001220050682]

[54]   Senadhira D, Zapata-Arias FJ, Gregorio GB, *et al.* Development of the first salt-tolerant rice cultivar through indica/indica anther culture. Field Crops Res 2002; 76(2-3): 103-10.
[http://dx.doi.org/10.1016/S0378-4290(02)00032-1]

[55]   Singh RK, Mishra B. Stable genotypes of rice for sodic soils. Ind J Genet 1997; 57(4): 431-8.

[56]   McCouch SR, Kochert G, Yu ZH, *et al.* Molecular mapping of rice chromosomes. Theor Appl Genet 1988; 76(6): 815-29.
[http://dx.doi.org/10.1007/BF00273666] [PMID: 24232389]

[57]   Causse MA, Fulton TM, Cho YG, *et al.* Saturated molecular map of the rice genome based on an interspecific backcross population. Genetics 1994; 138(4): 1251-74.
[http://dx.doi.org/10.1093/genetics/138.4.1251] [PMID: 7896104]

[58]   Kurata N, Moore G, Nagamura Y, *et al.* Conservation of genome structure between rice and wheat. Biotechnology (N Y) 1994; 12(3): 276-8.
[http://dx.doi.org/10.1038/nbt0394-276]

[59]   Harushima Y, Yano M, Shomura A, *et al.* A high-density rice genetic linkage map with 2275 markers using a single F2 population. Genetics 1998; 148(1): 479-94.
[http://dx.doi.org/10.1093/genetics/148.1.479] [PMID: 9475757]

[60]   Xiao J, Li J, Yuan L, Tanksley SD. Identification of QTLs affecting traits of agronomic importance in a recombinant inbred population derived from a subspecific rice cross. Theor Appl Genet 1996; 92(2): 230-44.
[http://dx.doi.org/10.1007/BF00223380] [PMID: 24166172]

[61]   Gregorio GB, Senadhira D, Mendoza RD, Manigbas NL, Roxas JP, Guerta CQ. Progress in breeding for salinity tolerance and associated abiotic stresses in rice. Field Crops Res 2002; 76(2-3): 91-101.
[http://dx.doi.org/10.1016/S0378-4290(02)00031-X]

[62]   Thomson MJ, de Ocampo M, Egdane J, *et al.* Characterizing the Saltol quantitative trait locus for salinity tolerance in rice. Rice (N Y) 2010; 3(2-3): 148-60.
[http://dx.doi.org/10.1007/s12284-010-9053-8]

[63]   Lin HX, Zhu MZ, Yano M, *et al.* QTLs for Na+ and K+ uptake of the shoots and roots controlling rice salt tolerance. Theor Appl Genet 2004; 108(2): 253-60.
[http://dx.doi.org/10.1007/s00122-003-1421-y] [PMID: 14513218]

[64]   Alam R, Sazzadur Rahman M, Seraj ZI, *et al.* Investigation of seedling-stage salinity tolerance QTLs using backcross lines derived from *Oryza sativa* L. Pokkali. Plant Breed 2011; 130(4): 430-7.
[http://dx.doi.org/10.1111/j.1439-0523.2010.01837.x]

[65]   Yu J, Zhao W, Tong W, *et al.* A genome-wide association study reveals candidate genes related to salt

tolerance in rice (*Oryza sativa*) at the germination stage. Int J Mol Sci 2018; 19(10): 3145.
[http://dx.doi.org/10.3390/ijms19103145] [PMID: 30322083]

[66] Wang Z, Wang J, Bao Y, Wu Y, Zhang H. Quantitative trait loci controlling rice seed germination under salt stress. Euphytica 2011; 178(3): 297-307.
[http://dx.doi.org/10.1007/s10681-010-0287-8]

[67] Prasad SR, Bagali PG, Hittalmani S, Shashidhar HE. Molecular mapping of quantitative trait loci associated with seedling tolerance to salt stress in rice (*Oryza sativa* L.). Curr Sci 2000; 162-4.

[68] Yao MZ, Wang JF, Chen HY, Zhai HQ, Zhang HS. Inheritance and QTL mapping of salt tolerance in rice. Rice Sci 2005; 12: 25-32.

[69] Egdane JA. QTL mapping for salinity tolerance in rice using Kalarata-Azucena population. Agris 2012; 37: 115.

[70] He Y, Yang B, He Y, *et al.* A quantitative trait locus, *QSE 3*, promotes seed germination and seedling establishment under salinity stress in rice. Plant J 2019; 97(6): 1089-104.
[http://dx.doi.org/10.1111/tpj.14181] [PMID: 30537381]

[71] Amoah NKA, Akromah R, Kena AW, Manneh B, Dieng I, Bimpong IK. Mapping QTLs for tolerance to salt stress at the early seedling stage in rice (*Oryza sativa* L.) using a newly identified donor 'Madina Koyo'. Euphytica 2020; 216(10): 156.
[http://dx.doi.org/10.1007/s10681-020-02689-5]

[72] Nakhla WR, Sun W, Fak Yang K, Zhang C, Yu S. Identification of QTLs for salt tolerance at the germination and seedling stages in Rice 2021; 10(10): 428.

[73] Zeng P, Zhu P, Qian L, *et al.* Identification and fine mapping of qGR6.2, a novel locus controlling rice seed germination under salt stress. BMC Plant Biol 2021; 21(1): 36.
[http://dx.doi.org/10.1186/s12870-020-02820-7] [PMID: 33422012]

[74] Lee SY, Ahn JH, Cha YS, *et al.* Mapping of quantitative trait loci for salt tolerance at the seedling stage in rice. Molecules & Cells (Springer Science & Business Media BV) 2006; 212: 192-6.

[75] Ammar MHM, Singh RK, Singh AK, Mohapatra T, Sharma TR, Singh NK. Mapping QTLs for salinity tolerance at seedling stage in rice *Oryza sativa* L. African Crop Science Proceedings 2007; 8: 617-20.

[76] Islam M, Salambr M, Hassanbr L, Collardbr B, Gregorio R. QTL mapping for salinity tolerance at seedling stage in rice. Emir J Food Agric 2011; 23(2): 137.
[http://dx.doi.org/10.9755/ejfa.v23i2.6348]

[77] Dahanayaka BA, Gimhani DR, Kottearachchi NS, Samarasighe WLG. Qtl mapping For Salinity Tolerance Using An Elite Rice (*Oryza sativa*) Breeding Population. SABRAO J Breed Genet 2017; 49(2): 123-34.

[78] Chen T, Zhu Y, Chen K, *et al.* Identification of new QTL for salt tolerance from rice variety Pokkali. J Agro Crop Sci 2020; pp. 1-12.

[79] Rahman MA, Bimpong IK, Bizimana JB, *et al.* Mapping QTLs using a novel source of salinity tolerance from Hasawi and their interaction with environments in rice. Rice (N Y) 2017; 10(1): 47.
[http://dx.doi.org/10.1186/s12284-017-0186-x] [PMID: 29098463]

[80] Yang T, Zhang S, Hu Y, *et al.* The role of a potassium transporter OsHAK5 in potassium acquisition and transport from roots to shoots in rice at low potassium supply levels. Plant Physiol 2014; 166(2): 945-59.
[http://dx.doi.org/10.1104/pp.114.246520] [PMID: 25157029]

[81] Flowers TJ, Koyama ML, Flowers SA, Sudhakar C, Singh KP, Yeo AR. QTL: their place in engineering tolerance of rice to salinity. J Exp Bot 2000; 51(342): 99-106.
[http://dx.doi.org/10.1093/jexbot/51.342.99] [PMID: 10938800]

[82] Koyama ML, Levesley A, Koebner RMD, Flowers TJ, Yeo AR. Quantitative trait loci for component

physiological traits determining salt tolerance in rice. Plant Physiol 2001; 125(1): 406-22.
[http://dx.doi.org/10.1104/pp.125.1.406] [PMID: 11154348]

[83]   Lang NT, Yanagihara S, Buu BC. A microsatellite marker for a gene conferring salt tolerance on rice at the vegetative and reproductive stages. SABRAO J Breed Genet 2001; 33: 1-10. a

[84]   Lang NT, Yanagihara S, Buu BC. QTL analysis of salt tolerance in rice (*Oryza sativa* L.). SABRAO J Breed Genet 2001; 33: 11-20. b

[85]   Sabouri H, Rezai AM, Moumeni A, Kavousi A, Katouzi M, Sabouri A. QTLs mapping of physiological traits related to salt tolerance in young rice seedlings. Biol Plant 2009; 53(4): 657-62.
[http://dx.doi.org/10.1007/s10535-009-0119-7]

[86]   Zang J, Sun Y, Wang Y, *et al*. Dissection of genetic overlap of salt tolerance QTLs at the seedling and tillering stages using backcross introgression lines in rice. Sci China C Life Sci 2008; 51(7): 583-91.
[http://dx.doi.org/10.1007/s11427-008-0081-1] [PMID: 18622741]

[87]   Ahmadi J, Mohammad HF. Identification and mapping of quantitative trait loci associated with salinity tolerance in rice using SSR markers. Iran J Biotechnol 2011; 9: 21-30.

[88]   Padmavathi G, Gautam RK. Ramesh K, Ramdeen, Gregorio GB, Ram T, Mishra B, Viraktamath BC. Mapping QTLs for alkalinity tolerance in a rice ultivar, CSR 27. Oral presentation in The 12th SABRAO Congress on Plant Breeding towards 2025- Challenges in a Rapidly Changing World, 201 Thailand during 13th to 16th January 2012.

[89]   Liang J, Qu Y, Yang C, *et al*. Identification of QTLs associated with salt or alkaline tolerance at the seedling stage in rice under salt or alkaline stress. Euphytica 2015; 201(3): 441-52.
[http://dx.doi.org/10.1007/s10681-014-1236-8]

[90]   Leon TB, Linscombe S, Subudhi PK. QTL Mapping for Salinity Tolerance in Rice (*Oryza sativa*). Using Ultra-High Density SNP Markers Rice 2016; 9: 52.

[91]   Puram VRR, Ontoy J, Subudhi PK. Identification of QTLs for Salt Tolerance Traits and Prebreeding Lines with Enhanced Salt Tolerance in an Introgression Line Population of Rice. Plant Mol Biol Report 2018; 36(5-6): 695-709.
[http://dx.doi.org/10.1007/s11105-018-1110-2]

[92]   Ammar MHM, Pandit A, Singh RK, *et al*. Mapping of QTLs controlling Na+, K+ and Cl− ion concentrations in salt tolerant indica rice variety CSR27. J Plant Biochem Biotechnol 2009; 18(2): 139-50.
[http://dx.doi.org/10.1007/BF03263312]

[93]   Pandit A, Rai V, Bal S, *et al*. Combining QTL mapping and transcriptome profiling of bulked RILs for identification of functional polymorphism for salt tolerance genes in rice (Oryza sativa L.). Mol Genet Genomics 2010; 284(2): 121-36.
[http://dx.doi.org/10.1007/s00438-010-0551-6] [PMID: 20602115]

[94]   Mohammadi R, Mendioro MS, Diaz GQ, Gregorio GB, Singh RK. Mapping quantitative trait loci associated with yield and yield components under reproductive stage salinity stress in rice (Oryza sativa L.). J Genet 2013; 92(3): 433-43.
[http://dx.doi.org/10.1007/s12041-013-0285-4] [PMID: 24371165]

[95]   Bimpong IK, Manneh B, Diop B, *et al*. New quantitative trait loci for enhancing adaptation to salinity in rice from Hasawi, a Saudi landrace into three African cultivars at the reproductive stage. Euphytica 2014; 200(1): 45-60.
[http://dx.doi.org/10.1007/s10681-014-1134-0]

[96]   Hossain H, Rahman MA, Alam MS, Singh RK. Mapping of Quantitative Trait Loci Associated with Reproductive-Stage Salt Tolerance in Rice. J Agron Crop Sci 2015; 201(1): 17-31.
[http://dx.doi.org/10.1111/jac.12086]

[97]   Tiwari S, Sl K, Kumar V, *et al*. Mapping QTLs for salt tolerance in rice (*Oryza sativa* L.) by bulked segregant analysis of recombinant inbred lines using 50K SNP chip. PLoS One 2016; 11(4):

e0153610.
[http://dx.doi.org/10.1371/journal.pone.0153610] [PMID: 27077373]

[98] Liu Y, Li C, Shi X, Feng H, Wang Y. Identification of QTLs with additive, epistatic, and QTL × environment interaction effects for the bolting trait in Brassica rapa L. Euphytica 2016; 210(3): 427-39.
[http://dx.doi.org/10.1007/s10681-016-1710-6]

[99] Mondal S, Borromeo TH, Diaz MGQ, *et al.* Dissecting QTLs for Reproductive Stage Salinity Tolerance in Rice from BRRI dhan 47. Plant Breed Biotechnol 2019; 7(4): 302-12.
[http://dx.doi.org/10.9787/PBB.2019.7.4.302]

[100] Chattopadhyay K, Mohanty SK, Vijayan J, *et al.* Genetic Dissection of Component Traits for Salinity Tolerance at Reproductive Stage in Rice. Plant Mol Biol Report 2020.
[http://dx.doi.org/10.1007/s11105-020-01257-4]

[101] Pundir P, Devi A, Krishnamurthy SL, Sharma PC, Vinaykumar NM. QTLs in salt rice variety CSR10 reveals salinity tolerance at reproductive stage. Acta Physiol Plant 2021; 43(2): 35.
[http://dx.doi.org/10.1007/s11738-020-03183-0]

[102] Tian L, Tan L, Liu F, Cai H, Sun C. Identification of quantitative trait loci associated with salt tolerance at seedling stage from Oryza rufipogon. J Genet Genomics 2011; 38(12): 593-601.
[http://dx.doi.org/10.1016/j.jgg.2011.11.005] [PMID: 22196402]

[103] Wang S, Cao M, Ma X, *et al.* Integrated RNA Sequencing and QTL Mapping to Identify Candidate Genes from *Oryza rufipogon* Associated with Salt Tolerance at the Seedling Stage. Front Plant Sci 2017; 8: 1427.
[http://dx.doi.org/10.3389/fpls.2017.01427] [PMID: 28861103]

[104] Yuan L, Zhang L, Wei X, *et al.* Quantitative Trait Locus Mapping of Salt Tolerance in Wild Rice *Oryza longistaminata.* Int J Mol Sci 2022; 23(4): 2379.
[http://dx.doi.org/10.3390/ijms23042379] [PMID: 35216499]

[105] Claes B, Dekeyser R, Villarroel R, *et al.* Characterization of a rice gene showing organ-specific expression in response to salt stress and drought. Plant Cell 1990; 2(1): 19-27.
[PMID: 2152105]

[106] Niones JM. Fine mapping of the salinity tolerance gene on chromosome 1 of rice (*Oryza sativa L.*), using near isogenic lines. MS dissertation. College, Laguna, Philippines: University of the Philippines Los Baños, Laguna 2004.

[107] Takehisa H, Ueda T, Fukuta Y, *et al.* Epistatic interaction of QTLs controlling leaf bronzing in rice (*Oryza sativa* L.) grown in a saline paddy field. Breed Sci 2006; 56(3): 287-93.
[http://dx.doi.org/10.1270/jsbbs.56.287]

[108] Sabouri H, Sabouri A. New evidence of QTLs attributed to salinity tolerance in rice. Afr J Biotechnol 2008; 7: 4376-83.

[109] Cheng HT, Jang H, Xue DW, *et al.* Mapping of QTLs underlying tolerance to alkali at germination and early seedling stages in rice. Zuo Wu Xue Bao 2009; 34(10): 1719-27. [acta Agronomica Sinica].
[http://dx.doi.org/10.3724/SP.J.1006.2008.01719]

[110] Kim DM, Ju HG, Kwon TR, Oh CS, Ahn SN. Mapping QTLs for salt tolerance in an introgression line population between Japonica cultivars in rice. J Crop Sci Biotechnol 2009; 12(3): 121-8.
[http://dx.doi.org/10.1007/s12892-009-0108-6]

[111] Javed MA, Huyop FZ, Wagiran A, Salleh FM. Identification of QTLs for morph-physiological traits related to salinity tolerance at seedling stage in indica rice. Procedia Environ Sci 2011; 8: 389-95.
[http://dx.doi.org/10.1016/j.proenv.2011.10.061]

[112] Wang Z, Cheng J, Chen Z, *et al.* Identification of QTLs with main, epistatic and QTL × environment interaction effects for salt tolerance in rice seedlings under different salinity conditions. Theor Appl Genet 2012; 125(4): 807-15.

[http://dx.doi.org/10.1007/s00122-012-1873-z] [PMID: 22678666]

[113]   Ghomi K, Rabiei B, Sabouri H, Sabouri A. Mapping QTLs for traits related to salinity tolerance at seedling stage of rice (Oryza sativa L.): an agrigenomics study of an Iranian rice population. OMICS 2013; 17(5): 242-51.
[http://dx.doi.org/10.1089/omi.2012.0097] [PMID: 23638881]

[114]   Mardani Z, Rabiei B, Sabouri H, Sabouri A. Identification of molecular markers linked to salt-tolerant genes at germination stage of rice. Plant Breed 2014; 133(2): 196-202.
[http://dx.doi.org/10.1111/pbr.12136]

[115]   Zheng H, Zhao H, Liu H, Wang J, Zou D. QTL analysis of Na+ and K+ concentrations in shoots and roots under NaCl stress based on linkage and association analysis in japonica rice. Euphytica 2015; 201(1): 109-21.
[http://dx.doi.org/10.1007/s10681-014-1192-3]

[116]   Calapit-Palao CD, Vina CB, Thomson MJ, Singh RK. QTL identification for reproductive-stage salinity tolerance in rice (*Oryza sativa* L.). Proceedings of SABRAO 13th Congress and international conference. September 14-16, 2015; Bogor, Indonesia.

[117]   Qiu X, Yuan Z, Liu H, *et al.* Identification of salt tolerance-improving quantitative trait loci alleles from a salt-susceptible rice breeding line by introgression breeding. Plant Breed 2015; 134(6): 653-60.
[http://dx.doi.org/10.1111/pbr.12321]

[118]   Gimhani DR, Gregorio GB, Kottearachchi NS, Samarasinghe WLG. SNP-based discovery of salinity-tolerant QTLs in a bi-parental population of rice (Oryza sativa). Mol Genet Genomics 2016; 291(6): 2081-99.
[http://dx.doi.org/10.1007/s00438-016-1241-9] [PMID: 27535768]

[119]   Fayed AM, Farid MA. Mapping of quantitative trait loci (QTL) for Na+ and K+ uptake controlling rice salt tolerance (*Oryza sativa* L.). Int J Curr Microbiol Appl Sci 2017; 6(1): 462-71.
[http://dx.doi.org/10.20546/ijcmas.2017.601.054]

[120]   De Leon TB, Linscombe S, Subudhi PK. Identification and validation of QTLs for seedling salinity tolerance in introgression lines of a salt tolerant rice landrace 'Pokkali'. PLoS One 2017; 12(4): e0175361.
[http://dx.doi.org/10.1371/journal.pone.0175361] [PMID: 28388633]

[121]   Rahman MA, Bimpong IK, Bizimana JB, *et al.* Mapping QTLs using a novel source of salinity tolerance from Hasawi and their interaction with environments in rice. Rice (N Y) 2017; 10(1): 47.
[http://dx.doi.org/10.1186/s12284-017-0186-x] [PMID: 29098463]

[122]   Bizimana JB, Luzi-Kihupi A, Murori RW, Singh RK. Identification of quantitative trait loci for salinity tolerance in rice (*Oryza sativa* L.) using IR29/Hasawi mapping population. J Genet 2017; 96(4): 571-82.
[http://dx.doi.org/10.1007/s12041-017-0803-x] [PMID: 28947705]

[123]   Puram VRR, Ontoy J, Linscombe S, Subudhi PK. Genetic dissection of seedling stage salinity tolerance in rice using introgression lines of a salt tolerant landrace Nona Bokra. J Hered 2017; 108(6): 658-70.
[http://dx.doi.org/10.1093/jhered/esx067] [PMID: 28821187]

[124]   Jahan N, Zhang Y, Lv Y, *et al.* QTL analysis for rice salinity tolerance and fine mapping of a candidate locus qSL7 for shoot length under salt stress. Plant Growth Regul 2020; 90(2): 307-19.
[http://dx.doi.org/10.1007/s10725-019-00566-3]

[125]   Lei L, Han Z, Cui B, *et al.* Mapping of a major QTL for salinity tolerance at the bud burst stage in rice (*Oryza sativa* L) using a high-density genetic map. Euphytica 2021; 217(8): 167.
[http://dx.doi.org/10.1007/s10681-021-02901-0]

[126]   Takehisa H, Shimodate T, Fukuta Y, *et al.* Identification of quantitative trait loci for plant growth of rice in paddy field flooded with salt water. Field Crops Res 2004; 89(1): 85-95.

[http://dx.doi.org/10.1016/j.fcr.2004.01.026]

[127] Linh LH, Linh TH, Xuan TD, Ham LH, Ismail AM, Khanh TD. Molecular Breeding to Improve Salt Tolerance of Rice (*Oryza sativa* L.) in the Red River Delta of Vietnam. Int J Plant Genomics 2012; 2012: 1-9.
[http://dx.doi.org/10.1155/2012/949038]

[128] Huyen LTN, Cuc LM, Ismail AM, Ham LH. Introgression the Salinity Tolerance QTLs *Saltol* into AS996, the Elite Rice Variety of Vietnam. Am J Plant Sci 2012; 3(7): 981-7.
[http://dx.doi.org/10.4236/ajps.2012.37116]

[129] Vu HTT, Le DD, Ismail AM, Le HH. Marker-assisted backcrossing (MABC) for improved salinity tolerance in rice (*Oryza sativa* L.) to cope with climate change in Vietnam. Aust J Crop Sci 2012; 6(12): 1649-54.

[130] Hoque ABMZ, Haque MA, Sarker MRA, Rahman MA. Marker-Assisted introgression of saltol locus into genetic background of brri dhan49. Int J Biosci 2015; 6(12): 71-80.
[http://dx.doi.org/10.12692/ijb/6.12.71-80]

[131] Usatov AV, Alabushev AV, Kostylev PI, Azarin KV, Makarenko MS, Usatova OA. Introgression the *SalTol* QTL into the Elite Rice Variety of Russia by Marker-Assisted Selection
. Am J Agric Biol Sci 2015; 10(4): 165-9.
[http://dx.doi.org/10.3844/ajabssp.2015.165.169]

[132] Babu NN, Krishnan SG, Vinod KK, *et al.* Marker Aided Incorporation of Saltol, a Major QTL Associated with Seedling Stage Salt Tolerance, into Oryza sativa 'Pusa Basmati 1121'. Front Plant Sci 2017; 8: 41.
[http://dx.doi.org/10.3389/fpls.2017.00041] [PMID: 28184228]

[133] Rana MM, Takamatsu T, Baslam M, *et al.* Salt tolerance improvement in rice through efficient SNP marker-assisted selection coupled with speed-breeding. Int J Mol Sci 2019; 20(10): 2585-607.
[http://dx.doi.org/10.3390/ijms20102585] [PMID: 31130712]

[134] Borsani O, Valpuesta V, Botella MA. Developing salt tolerant plants in a new century: a molecular biology approach. Plant Cell Tissue Organ Cult 2003; 73(2): 101-15.
[http://dx.doi.org/10.1023/A:1022849200433]

[135] Rhodes D, Hanson AD. Quaternary ammonium and tertiary sulfoniu compounds in higher plants. Annu Rev Plant Physiol Plant Mol Biol 1993; 44(1): 357-84.
[http://dx.doi.org/10.1146/annurev.pp.44.060193.002041]

[136] Gao JP, Chao DY, Lin HX. Understanding abiotic stress tolerance mechanisms: recent studies on stress response in rice. J Integr Plant Biol 2007; 49(6): 742-50.
[http://dx.doi.org/10.1111/j.1744-7909.2007.00495.x]

[137] Sakamoto A, Murata AN, Murata N. Metabolic engineering of rice leading to biosynthesis of glycinebetaine and tolerance to salt and cold. Plant Mol Biol 1998; 38(6): 1011-9.
[http://dx.doi.org/10.1023/A:1006095015717] [PMID: 9869407]

[138] Su J, Chen PL, Wu R. Transgene expression of mannitol- 1-phosphate dehydrogenase enhanced the salt stress tolerance of the transgenic rice seedlings. Zhongguo Nong Ye Ke Xue 1999; 32: 101-3.

[139] Kishitani S, Takanami T, Suzuki M, *et al.* Compatibility of glycinebetaine in rice plants: evaluation using transgenic rice plants with a gene for peroxisomal betaine aldehyde dehydrogenase from barley. Plant Cell Environ 2000; 23(1): 107-14.
[http://dx.doi.org/10.1046/j.1365-3040.2000.00527.x]

[140] Shirasawa K, Takabe T, Takabe T, Kishitani S. Accumulation of glycinebetaine in rice plants that overexpress choline monooxygenase from spinach and evaluation of their tolerance to abiotic stress. Ann Bot (Lond) 2006; 98(3): 565-71.
[http://dx.doi.org/10.1093/aob/mcl126] [PMID: 16790464]

[141] Garg AK, Kim JK, Owens TG, *et al.* Trehalose accumulation in rice plants confers high tolerance

levels to different abiotic stresses. Proc Natl Acad Sci USA 2002; 99(25): 15898-903.
[http://dx.doi.org/10.1073/pnas.252637799] [PMID: 12456878]

[142]   Jang IC, Oh SJ, Seo JS, *et al*. Expression of a bifunctional fusion of the Escherichia coli genes for trehalose-6-phosphate synthase and trehalose-6-phosphate phosphatase in transgenic rice plants increases trehalose accumulation and abiotic stress tolerance without stunting growth. Plant Physiol 2003; 131(2): 516-24.
[http://dx.doi.org/10.1104/pp.007237] [PMID: 12586876]

[143]   Li HW, Zang BS, Deng XW, Wang XP. Overexpression of the trehalose-6-phosphate synthase gene *OsTPS1* enhances abiotic stress tolerance in rice. Planta 2011; 234(5): 1007-18.
[http://dx.doi.org/10.1007/s00425-011-1458-0] [PMID: 21698458]

[144]   Sahoo RK, Ansari MW, Tuteja R, Tuteja N. OsSUV3 transgenic rice maintains higher endogenous levels of plant hormones that mitigates adverse effects of salinity and sustains crop productivity. Rice (N Y) 2014; 7(1): 17.
[http://dx.doi.org/10.1186/s12284-014-0017-2] [PMID: 24383761]

[145]   Liao YD, Lin KH, Chen CC, Chiang CM. *Oryza sativa* protein phosphatase 1a (*OsPP1a*) involved in salt stress tolerance in transgenic rice. Mol Breed 2016; 36(3): 22.
[http://dx.doi.org/10.1007/s11032-016-0446-2]

[146]   Guo W, Chen T, Hussain N, Zhang G, Jiang L. Characterization of salinity tolerance of transgenic rice lines harboring HsCBL8 of wild barley (*Hordeum spontanum*) line from Qinghai-Tibet plateau. Front Plant Sci 2016; 7: 1678.
[http://dx.doi.org/10.3389/fpls.2016.01678] [PMID: 27891136]

[147]   Xiong L, Zhu JK. Molecular and genetic aspects of plant responses to osmotic stress. Plant Cell Environ 2002; 25(2): 131-9.
[http://dx.doi.org/10.1046/j.1365-3040.2002.00782.x] [PMID: 11841658]

[148]   Xu D, Duan X, Wang B, Hong B, Ho T, Wu R. Expression of a late embryogenesis abundance protein gene, *HVA1*, from barley confers tolerance to water deficit and salt stress in transgenic rice. Plant Physiol 1996; 110(1): 249-57.
[http://dx.doi.org/10.1104/pp.110.1.249] [PMID: 12226181]

[149]   Rohila JS, Jain RK, Wu R. Genetic improvement of Basmati rice for salt and drought tolerance by regulated expression of a barley *Hva1* cDNA. Plant Sci 2002; 163(3): 525-32.
[http://dx.doi.org/10.1016/S0168-9452(02)00155-3]

[150]   Hu TZ. OsLEA3, a late embryogenesis abundant protein gene from rice, confers tolerance to water deficit and salt stress to transgenic rice. Russ J Plant Physiol 2008; 55(4): 530-7. a
[http://dx.doi.org/10.1134/S1021443708040158]

[151]   Hong B, Barg R, Ho TD. Developmental and organ-specific expression of an ABA- and stress-induced protein in barley. Plant Mol Biol 1992; 18(4): 663-74.
[http://dx.doi.org/10.1007/BF00020009] [PMID: 1532749]

[152]   Chandra Babu R, Zhang J, Blum A, David Ho T-H, Wu R, Nguyen HT. HVA1, a LEA gene from barley confers dehydration tolerance in transgenic rice (*Oryza sativa* L.) *via* cell membrane protection. Plant Sci 2004; 166(4): 855-62.
[http://dx.doi.org/10.1016/j.plantsci.2003.11.023]

[153]   Cheng Z, Targolli J, Huang X, Wu R. Wu. Wheat LEA genes, *PMA80* and *PMA1959*, enhance dehydration tolerance of transgenic rice (*Oryza sativa* L.). Mol Breed 2002; 10(1/2): 71-82.
[http://dx.doi.org/10.1023/A:1020329401191]

[154]   Ganguly M, Datta K, Roychoudhury A, Gayen D, Sengupta DN, Datta SK. Overexpression of *Rab16A* gene in indica rice variety for generating enhanced salt tolerance. Plant Signal Behav 2012; 7(4): 502-9.
[http://dx.doi.org/10.4161/psb.19646] [PMID: 22499169]

[155]   Peng X, Ding X, Chang T, *et al.* Overexpression of a Vesicle Trafficking Gene, OsRab7, enhances salt tolerance in rice. ScientificWorldJournal 2014; 2014: 1-7.
[http://dx.doi.org/10.1155/2014/483526] [PMID: 24688390]

[156]   Kasuga M, Liu Q, Miura S, Yamaguchi-Shinozaki K, Shinozaki K. Improving plant drought, salt, and freezing tolerance by gene transfer of a single stress-inducible transcription factor. Nat Biotechnol 1999; 17(3): 287-91.
[http://dx.doi.org/10.1038/7036] [PMID: 10096298]

[157]   Saijo Y, Hata S, Kyozuka J, Shimamoto K, Izui K. Over-expression of a single Ca^{2+}-dependent protein kinase confers both cold and salt/drought tolerance on rice plants. Plant J 2000; 23(3): 319-27.
[http://dx.doi.org/10.1046/j.1365-313x.2000.00787.x] [PMID: 10929125]

[158]   Wang L, Cai H, Bai X, Li LW, Li Y, Zhu YM. Cultivation of transgenic rice plants with *OsCDPK7* gene and its salt tolerance. Yi Chuan 2008; 30(8): 1051-5.
[http://dx.doi.org/10.3724/SP.J.1005.2008.01051] [PMID: 18779157]

[159]   Shi H, Ishitani M, Kim C, Zhu JK. The *Arabidopsis thaliana* salt tolerance gene *SOS1* encodes a putative Na$^+$/H$^+$ antiporter. Proc Natl Acad Sci USA 2000; 97(12): 6896-901.
[http://dx.doi.org/10.1073/pnas.120170197] [PMID: 10823923]

[160]   Aharon GS, Apse MP, Duan S, Hua X, Blumwald E. Characterization of a family of vacuolar Na$^+$/H$^+$ antiporters in Arabidopsis thaliana. Plant Soil 2003; 253(1): 245-56.
[http://dx.doi.org/10.1023/A:1024577205697]

[161]   Apse MP, Aharon GS, Snedden WA, Blumwald E. Salt tolerance conferred by overexpression of a vacuolar Na+/H+ antiport in Arabidopsis. Science 1999; 285(5431): 1256-8.
[http://dx.doi.org/10.1126/science.285.5431.1256] [PMID: 10455050]

[162]   Ohta M, Hayashi Y, Nakashima A, *et al.* Introduction of a Na$^+$/H$^+$ antiporter gene from *Atriplex gmelini* confers salt tolerance to rice. FEBS Lett 2002; 532(3): 279-82.
[http://dx.doi.org/10.1016/S0014-5793(02)03679-7] [PMID: 12482579]

[163]   Fukuda A, Nakamura A, Tagiri A, *et al.* Function, intracellular localization and the importance in salt tolerance of a vacuolar Na(+)/H(+) antiporter from rice. Plant Cell Physiol 2004; 45(2): 146-59.
[http://dx.doi.org/10.1093/pcp/pch014] [PMID: 14988485]

[164]   Zhao F, Guo S, Zhang H, Zhao Y. Expression of yeast *SOD2* in transgenic rice results in increased salt tolerance. Plant Sci 2006; 170(2): 216-24.
[http://dx.doi.org/10.1016/j.plantsci.2005.08.017]

[165]   Hoang TML, Moghaddam L, Williams B, Khanna H, Dale J, Mundree SG. Development of salinity tolerance in rice by constitutive-overexpression of genes involved in the regulation of programmed cell death. Front Plant Sci 2015; 6: 175.
[http://dx.doi.org/10.3389/fpls.2015.00175] [PMID: 25870602]

[166]   Amin USM, Biswas S, Elias SM, *et al.* Enhanced salt tolerance conferred by the complete 2.3 kb cDNA of the rice vacuolar Na+/H+ antiporter gene compared to 1.9 kb coding region with 5′ UTR in transgenic lines of rice. Front Plant Sci 2016; 7: 14.
[http://dx.doi.org/10.3389/fpls.2016.00014] [PMID: 26834778]

[167]   Xiang Y, Huang Y, Xiong L. Characterization of stress-responsive *CIPK* genes in rice for stress tolerance improvement. Plant Physiol 2007; 144(3): 1416-28.
[http://dx.doi.org/10.1104/pp.107.101295] [PMID: 17535819]

[168]   Diédhiou CJ, Popova OV, Dietz KJ, Golldack D. The SNF1-type serine-threonine protein kinase SAPK4regulates stress-responsive gene expression in rice. BMC Plant Biol 2008; 8(1): 49.
[http://dx.doi.org/10.1186/1471-2229-8-49] [PMID: 18442365]

[169]   Ito Y, Katsura K, Maruyama K, *et al.* Functional analysis of rice DREB1/CBF-type transcription factors involved in cold-responsive gene expression in transgenic rice. Plant Cell Physiol 2006; 47(1): 141-53.

[http://dx.doi.org/10.1093/pcp/pci230] [PMID: 16284406]

[170] Wang Q, Guan Y, Wu Y, Chen H, Chen F, Chu C. Overexpression of a rice *OsDREB1F* gene increases salt, drought, and low temperature tolerance in both Arabidopsis and rice. Plant Mol Biol 2008; 67(6): 589-602.
[http://dx.doi.org/10.1007/s11103-008-9340-6] [PMID: 18470484]

[171] Hou X, Xie K, Yao J, Qi Z, Xiong L. A homolog of human ski-interacting protein in rice positively regulates cell viability and stress tolerance. Proc Natl Acad Sci USA 2009; 106(15): 6410-5.
[http://dx.doi.org/10.1073/pnas.0901940106] [PMID: 19339499]

[172] Parvin S, Biswas S, Razzaque S, *et al.* Salinity and drought tolerance conferred by in planta transformation of SNAC1 transcription factor into a high-yielding rice variety of Bangladesh. Acta Physiol Plant 2015; 37(4): 68.
[http://dx.doi.org/10.1007/s11738-015-1817-8]

[173] Tanaka Y, Hibino T, Hayashi Y, *et al.* Salt tolerance of transgenic rice overexpressing yeast mitochondrial Mn-SOD in chloroplasts. Plant Sci 1999; 148(2): 131-8.
[http://dx.doi.org/10.1016/S0168-9452(99)00133-8]

[174] Zhao F, Wang Z, Zhang Q, Zhao Y, Zhang H. RETRACTED ARTICLE: Analysis of the physiological mechanism of salt-tolerant transgenic rice carrying a vacuolar Na+/H+ antiporter gene from Suaeda salsa. J Plant Res 2006; 119(2): 95-104.
[http://dx.doi.org/10.1007/s10265-005-0250-2] [PMID: 16565882]

[175] Nagamiya K, Motohashi T, Nakao K, *et al.* Enhancement of salt tolerance in transgenic rice expressing an *Escherichia coli* catalase gene, katE. Plant Biotechnol Rep 2007; 1(1): 49-55.
[http://dx.doi.org/10.1007/s11816-007-0007-6]

[176] Prashanth SR, Sadhasivam V, Parida A. Over expression of cytosolic copper/zinc superoxide dismutase from a mangrove plant *Avicennia marina* in *indica* Rice var Pusa Basmati-1 confers abiotic stress tolerance. Transgenic Res 2008; 17(2): 281-91.
[http://dx.doi.org/10.1007/s11248-007-9099-6] [PMID: 17541718]

[177] Hoshida H, Tanaka Y, Hibino T, *et al.* Enhanced tolerance to salt stress in transgenic rice that overexpresses chloroplast glutamine synthetase. Plant Mol Biol 2000; 43(1): 103-11.
[http://dx.doi.org/10.1023/A:1006408712416] [PMID: 10949377]

[178] Serrano R, Mulet JM, Rios G, *et al.* A glimpse of the mechanisms of ion homeostasis during salt stress. J Exp Bot 1999; 50(Special_Issue): 1023-36.
[http://dx.doi.org/10.1093/jxb/50.Special_Issue.1023]

[179] Garciadeblás B, Senn ME, Bañuelos MA, Rodríguez-Navarro A. Sodium transport and HKT transporters: the rice model. Plant J 2003; 34(6): 788-801.
[http://dx.doi.org/10.1046/j.1365-313X.2003.01764.x] [PMID: 12795699]

[180] Obata T, Kitamoto HK, Nakamura A, Fukuda A, Tanaka Y. Rice shaker potassium channel *OsKAT1* confers tolerance to salinity stress on yeast and rice cells. Plant Physiol 2007; 144(4): 1978-85.
[http://dx.doi.org/10.1104/pp.107.101154] [PMID: 17586689]

[181] Majee M, Maitra S, Dastidar KG, *et al.* A novel salt-tolerant L-myo-inositol-1-phosphate synthase from *Porteresia coarctata* (Roxb.) Tateoka, a halophytic wild rice: molecular cloning, bacterial overexpression, characterization, and functional introgression into tobacco-conferring salt tolerance phenotype. J Biol Chem 2004; 279(27): 28539-52.
[http://dx.doi.org/10.1074/jbc.M310138200] [PMID: 15016817]

[182] Mishra NS, Pham ZH, Sopory SK, Tuteja N. Pea DNA helicase 45 overexpression in tobacco confers high salinity tolerance without affecting yield 2005.

[183] Nath M, Yadav S, Kumar Sahoo R, Passricha N, Tuteja R, Tuteja N. PDH45 transgenic rice maintain cell viability through lower accumulation of Na+, ROS and calcium homeostasis in roots under salinity stress. J Plant Physiol 2016; 191: 1-11.

[http://dx.doi.org/10.1016/j.jplph.2015.11.008] [PMID: 26687010]

[184] Mukherjee R, Mukherjee A, Bandyopadhyay S, *et al.* Selective manipulation of the inositol metabolic pathway for induction of salt-tolerance in indica rice variety. Sci Rep 2019; 9(1): 5358.
[http://dx.doi.org/10.1038/s41598-019-41809-7] [PMID: 30926863]

[185] Joshi R, Sahoo KK, Singh AK, *et al.* Enhancing trehalose biosynthesis improves yield potential in marker-free transgenic rice under drought, saline, and sodic conditions. J Exp Bot 2020; 71(2): 653-68.
[http://dx.doi.org/10.1093/jxb/erz462] [PMID: 31626290]

[186] Awaji SM, Prashantkumar S. Hanjagi, Pushpa BN, Sashidhar VR. Over-expression of plasma membrane Na+ /H+ antiporter *OsSOS1* gene improves salt tolerance in transgenic rice plants. Oryza (Cuttack) 2020; 57(4): 277-87.
[http://dx.doi.org/10.35709/ory.2020.57.4.3]

[187] Fan Y, Xie J, Zhang F. Overexpression of miR5505 enhanced drought and salt resistance in rice (*Oryza sativa*) 2022.
[http://dx.doi.org/10.1101/2022.01.13.476146]

[188] Kim S, Park SI, Kwon H, *et al.* The Rice Abscisic Acid-Responsive RING Finger E3 Ligase *OsRF1* Targets *OsPP2C09* for Degradation and Confers Drought and Salinity Tolerance in Rice. Front Plant Sci 2022; 12: 797940.
[http://dx.doi.org/10.3389/fpls.2021.797940] [PMID: 35095969]

[189] Xu C, Luo M, Sun X, *et al. SiMYB19* from Foxtail Millet (*Setaria italica*) Confers Transgenic Rice Tolerance to High Salt Stress in the Field. Int J Mol Sci 2022; 23(2): 756.
[http://dx.doi.org/10.3390/ijms23020756] [PMID: 35054940]

CHAPTER 5

Morphological and Physiological Responses of Plants Under Temperature Stress and Underlying Mechanisms

Asma Shakeel[1,*], **Syed Andleeba Jan**[1], **Shakeel A Mir**[2], **Z. Mehdi**[1], **Inayat M. Khan**[1] and **Mehnaz Shakeel**[1]

[1] *Faculty of Agriculture, SKUAST, Kashmir, 193201, India*
[2] *Faculty of Horticulture, SKUAST, Kashmir 190025, India*

Abstract: During evolution, plants are exposed to a wide range of beneficial and detrimental environmental conditions. Among these, temperature stress could retard plant growth and development, and even threaten survival. In agriculture, due to temperature stress, crop yield might be reduced remarkably and consequently damage food security. Fortunately, to mitigate these losses, plants have evolved various mechanisms for adaptation, avoidance and acclimatization to overcome temperature stress. For example, chilling or freezing injury can lead to the disruption of many physiological processes in plants, *e.g.*, water status, photosynthesis, respiration, and even most of the metabolism, and thus, various adaptive mechanisms could be activated in plants to avoid damage by the ice crystal formation or other chilling damages. These temperature-stress-tolerant mechanisms for high-temperature stress, cold stress, chilling injury, and freezing injury have been intensively revealed by researchers, and this present chapter attempts to summarize them systematically.

Keywords: Cold Stress, Chilling Injury, Cold Tolerance, Freezing Injury, High-Temperature Stress, Heat Tolerance.

1. INTRODUCTION

Plants have a prerequisite of some abiotic factors, including light, temperature, air, water, and some chemical factors, such as organic and inorganic nutrients, which are balanced and accurate at the same time to optimize growth and development. All deviations from the optimal levels of these essential factors for plant growth may cause impacts in the standard functions of the plant giving rise to abiotic stress. Theoretically, abiotic stress can be either elastic (reversible) or plastic (irreversible). Temperature stress is designated as to the metabolic proc-

---

* **Corresponding author Asma Shakeel:** Faculty of Agriculture, SKUAST, Kashmir, 193201, India; Tel: +918082081151 E-mail: asmasm1319@skuastkashmir.ac.in

Jen-Tsung Chen (Ed.)

-esses in plants plastic biological stress, as the functions of plants do not return to normal under temperature stress. Temperature is one of the most important biological factors that influence the natural spread of plants. This factor also influences the rise and fall of agricultural crop yields. Temperature is an important abiotic factor for plant photosynthesis, respiration, and transpiration. In various ecosystems, plants differ in their temperature tolerance, which ranges from below 0°C to 60°C. The fluctuation in temperature under different habitats leads to damage to the metabolic processes in plants.

2. TEMPERATURE STRESS

Earth's temperature changes along longitudinal and altitudinal lines and also with the season. Most plants are specially adapted to function over a particular range of temperatures. Fluctuation in temperature above or below this range limits the survival of plants. Climate change is leading plant scientists to increasingly be concerned about temperature stress, mainly due to the potential impact of agriculture [1]. High temperatures caused by global climate change are now considered one of the main abiotic constraints restricting agricultural production [2]. The mean global temperature is now predicted to be 1.7-4.9 °C warmer than the current level by 2,100 due to the impact of the growth of the human population [3]. This prediction is raising scientific concern because temperature stress is known to affect the life cycle of organisms. As a sessile organism, plants cannot demonstrate a movement to favorable conditions. Consequently, gets exposed to lethal conditions due to High Temperature (HT) stress.

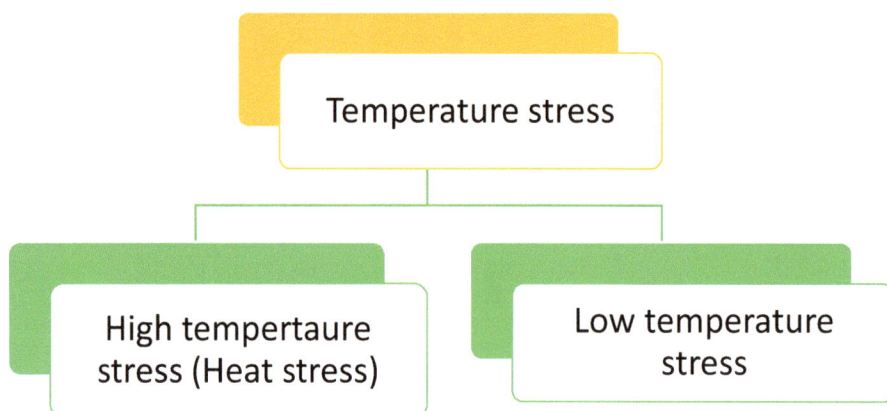

Temperature stress

High tempertaure stress (Heat stress)

Low temperature stress

3. PLANT RESPONSES TO HIGH TEMPERATURE (HT) STRESS: AN OVERVIEW

High temperature is an environmental stress that occurs due to a hike in temperature for a long period beyond the captivity threshold, causing irreparable damage to the growth and development of plants [4]. The growth and development of plants cover a large number of biological, chemical reactions and sensitivity to high temperatures. Heat stress affects almost all plant processes, from germination to yield [5 - 8]. Severe cellular damage or even cell death occurs within minutes under high temperatures, which leads to the fatal collapse of cellular association [9]. Damage or death caused by HT includes protein degeneration, increased membrane lipid fluidity, and changes in enzyme reaction efficiency, resulting in metabolic imbalances.

3.1. Germination Stage

Germination is the first step of plant development to be impacted. As previously stated, several plant species experience a drop in germination percentage as a result of HT stress. High temperatures also damage plant germination, seedling vigor, and radicle and plumule development [10 - 12]. HT affects seed germination differently depending on the plant species. Heat stress harms a variety of crops during seed germination, albeit the temperature ranges differ per crop type. Heat stress has a variety of negative effects on geminated seedling germination percentage, plant emergence, seedling vigor, radicle and plumule development. Seed germination is also inhibited at high temperatures, which is caused by the induction of ABA. Under HT, Cell size is reduced due to the loss of cell content, and ultimately, growth is affected. Another reason for the relative growth rate (RGR) decline is the reduction in the net assimilation rate (NAR) of HT in corn, wheat, and sugarcane.

3.2. Photosynthesis

One of the heat-sensitive physiological processes in plants is photosynthesis. The photosynthetic capability of C3 plants is significantly higher than that of C4 plants when temperatures are high [13]. Under the influence of HT, the carbon metabolism of stroma and the photochemical reactions of thylakoid lamellae are the main locations of damage to chloroplasts [14]. The thylakoid membrane is extremely vulnerable to HT. Under HT, considerable changes in chloroplasts occur, including thylakoids' structural organization, grana stacking loss and grana inflammation. Photosystem II activity is greatly decreased, if not stopped, at high temperatures. The number of photosynthetic pigments is likewise reduced by HT. Heat resistance is directly linked to plants' ability to stimulate the exchange of leaf gases and the rate of CO_2 assimilation when temperatures are high. The state of the leaf water, the transmission of the leaf stomata and the concentration of CO_2 between the cells are affected by heat stress. Closure of the stomata Photosynthesis is further hampered by high temperatures, which affects intercellular CO_2. Heat stress induces lipid peroxidation in the chloroplast and thylakoid membranes, lowering chlorophyll pigment levels. Under heat stress, variables such as reduced soluble proteins, Rubisco-linked proteins (RBPs), large and small Rubisco-linked proteins in darkness, and increased light impede photosynthesis. Heat stress affects the activities of sucrose phosphate synthesis enzymes, ADP glucose pyrophosphorylase and invertase, which are all involved in starch and sugar production. Heat stress affects the potential of watering the leaves, the area of the leaves, and the senescence of pre-mature leaves, which have adverse effects on the overall photosynthetic rate of the plant. Under prolonged heat stress, glucose supplies are depleted, and plants become starved.

3.3. Reproductive Growth

Thermal stress affects all plant tissues, but reproductive tissues are the most sensitive; a rise in temperatures during the blooming period may fail the whole grain production cycle. Short periods of heat stress in the reproductive phase can cause a significant decrease in flower buds and flower abortion; although there are great differences in the sensitivity of flowers and flower varieties, there are also high variations in the sensitivity of plants and varieties [15]. At the reproductive stage, plants that do not produce flowers or fruit or seed at high temperatures cannot produce flowers [16, 17]. Impairment of male and female meiosis, diseases of pollen germination and growth of pollen tubes decreased reproductive viability of ovaries, the abnormality of the position of the stigma and the style, the lack of pollen grains preserved by the stigma, and the disruption of fertilization processes, Impediment to the growth of endosperms, proembryos, and unfertilized embryos [18], and significantly reduced anther dehiscence at heading stage all contribute to rising sterility under abiotic stress conditions [19, 20]. Rice spikelet sterility is increased (by 61 percent compared to control) by high night temperatures (32 °C). This is due to reduced pollen germination (36 percent) [21]. Excessive ethylene synthesis is caused by high stress, resulting in male sterility in rice pollens. Ethylene inhibits important enzymes in sugar–starch metabolism, weakening sink strength, causing grain abortion, and eventually resulting in sterile grain. High-temperature stress induces abscission and abortion of flowers, pods, and seeds, resulting in a decrease in soybean seed count [22].

3.4. Transpiration

The loss of water in the form of vapors from plants is called transpiration. It is a physical process that is controlled both by physical as well as physiological factors. The transpiration mechanism is critical for water transportation from the root to the shoot and leaf *via* mass flow inside a vascular bundle. Under thermal stress, the risk of damage caused to HT and water shortages in plants is dramatically increasing. Transpiration effectively protects plants by cooling the leaves and other green parts of plants through the evaporation of high-temperature stomas water. Exposure to HT (28 °C), *Arabidopsis* plants showed a higher evaporation rate and a higher leaf cooling capacity, improving plant adaptation to high temperatures [23]. The heat stress (38 °C for 4 days) in tomato plants remarkably builds up the transpiration rate. The transpiration rate declines after three days [24]. Transpiration not only shows a cooling effect in the leaf but also in the entire plant. When the environmental temperature was 63°C, *Pinus ponderosa* seedlings perished. However, few seedlings survived because their base stem temperatures were 15 degrees Celsius lower than the ambient air temperature. This drop in temperature is due to the increase in stomatal transmission, transpiration rate and faster water transport, which preserves the stems of leaves and seeds to provide a cooling impact to plants through the HT transmitting mechanism [25]. In contrast, the hydraulic conductivity of tomatoes was lowered by HT, resulting in a reduction in water absorption [26].

3.5. Water Relation

One of the most important components of the plant is the water status of the plant, which should be asserted under fluctuating temperatures; otherwise, environmental stress gravely affects the physiological processes [27]. Even though plants tend to maintain water balance under stressful circumstances, imbalance commonly occurs as a result of changes in water relations between plant tissue and the culture substrate or between plant components under various stress situations, such as HT stress. Also, the imbalance between symplastic and apoplastic phloem loading is induced by high-temperature stress. This might be one of the reasons for reduced water transportation. It may also result in a decrease in photosynthate transit, lowering carbohydrate buildup in pollen grains and stigmatic tissue [28]. HT influences plant root hydraulic movements and water conditions even when water supplies, relative humidity, and soil conditions are optimal [29]. Water-use efficiency is decreased under high temperatures in wheat [30]. During the reproductive stage, a reduction of water and nutrient supply occurs, resulting in poor fruit sets in tropical plants [31]. The possibility of reduced leaf water of Lotus critics is greatly reduced by the elevation of night temperature [32]. In tomatoes, hydraulic conductivity is reduced drastically under HT, causing a significant decrease in the water status of the plant due to declined stomatal movement [33]. In *Phaseolus vulgaris*, the water relation is impeded under HT stress which severely reduces growth [34]. Vapour pressure differences between intercellular air spaces and rooted environment absorption are essential for water transpiration to meet growth requirements. Short-term imbalances are buffered by photosynthate or nutrient storage, which stabilizes these inputs over time. Temperature changes affect the mechanism of material translocation and partitioning in a plant [35, 36]. Due to their low water content, seeds and spores are particularly heat resistant, and dry seeds exposed to extremely high temperatures for several hours can even induce a larger percentage of germination.

3.6. Oxidative Stress

Enzymes that control various metabolic pathways are fragile to high temperatures. Due to the dissociation of enzymes and metabolic pathways that may be caused by heat stress, undesirable and Inactive oxygen species (ROS) that cause oxidative damage, such as single oxygen (O_2), superoxide radicals (O_2), hydrogen peroxides (H_2O_2) and hydrogen radicals (OH-), accumulate [37]. ROS formation takes place primarily in PSI and PSII reaction centers in chloroplasts, but it also happens in other organelles, such as peroxisomes and mitochondria. Due to heat damage to photosystems, the absorption of photons is less. If photosystems absorb more photons than they need for CO_2 assimilation, the excess is called surplus electrons, which serve as a source of ROS. During mitochondrial ETCs processes, photooxidation reactions in chloroplasts through the Mehler reaction lead to the generation of O_2. The generation of hydroxyl radicals is caused by the interactions of H_2O_2 with O_2- (Haber-Weiss reaction), H_2O_2 with Fe^{2+} (Fenton reaction) and O_3 breakdown in the apoplastic region [38]. Even though ROS have a significant negative influence on plant metabolic processes, it has been proposed that ROS, through their signaling activity, activates the response of heat shocks, thereby enabling plant temperature tolerance.

3.7. Yield

The yield is the result of a variety of factors, such as the number of plants that germinate and grow, the production of dry matter, the number of seeds, and the size of the seeds. High temperature influences the crop, even a small increase in temperature (1.5 °C) has a remarkable negative effect on crop yield. In many cultivable crops, including cereals, pulses, and oilseeds, yield reductions have been reported due to high temperatures. Due to a reduction in photosynthesis, high temperatures restrict biomass output. Heat stress causes a decrease in assimilatory rate due to lower photosynthesis caused by membrane instability, increased maintenance respiration costs and a decrease in radiation use efficiency.

4. MITIGATION STRATEGIES FOR HIGH-TEMPERATURE STRESS

Survival under high temperatures can be achieved by various adaptation mechanisms. The adaptation of plants to heat stress includes mechanisms for avoiding and tolerating it, which include some approaches.

AVOIDANCE	• Changing leaf orientation • Transpirational cooling • Leaf rolling • Early maturity
TOLERANCE	• Alteration of membrane lipid components • Antioxidant defense • Expression of stress proteins • Osmoprotectants

Moreover, some genetic and molecular approaches are also signaled to counteract heat stress.

4.1. The Function of Modified Membrane in Heat Tolerance

Lipid peroxidation is a process of mutilation that occurs in living organisms. At high temperatures, membrane changes are usually caused by changes in membrane fluid. Membrane-based processes include three specifications, *i.e.*, plasmalemma, photosynthetic membranes, and mitochondrial membranes. One of the important considerations in high-temperature tolerance is membrane lipid saturation. Due to high temperature, the fluidity of the membrane increases, which leads to the putrefaction of the lipid membrane. It has been recognized that products of lipid peroxidation are produced from polyunsaturated precursors that involve small fragments of hydrocarbons such as ketones, malondialdehyde (MDA), and compounds related to them [39]. In cell membranes and organelles, lipid peroxidation occurs, which is the reactive oxygen species (ROS). The level is reached above the threshold, creating damage to normal cellular functioning [40]. The lipid peroxidation mechanism includes three stages: the first, the next, and the last stages. The initial step is the reaction of a free activated atom such as single oxygen ($^1O_2$, O_2^-, or OH^-), the substrate of the lipid, and produces a very reactive carbon-based radical lipid. Progress is a process of rapidly adding molecular oxygen to produce lipid peroxide radicals. Lipid peroxide radicals eliminate hydrogen atoms from other lipid molecules and produce lipid hydroperoxides and other extremely reactive carbon-centered radicals, which then elongates the chain reaction, and the final phase of the end includes the end of the peroxidation of lipids that occurs by coupling any two radicals to form non-radical products. These products are stable, but they cannot propagate the lipid peroxidation reaction. Transitional metal ions, such as copper and iron, are essential for lipid peroxidation. In membrane unification, the function of membrane proteins is sensitive to high temperatures because it causes changes in the tertiary and fourth structures of membrane proteins. Both temperatures, upshift and downshift, cause the unfolding of membrane proteins.

4.2. The Function of Antioxidative Defense in Heat Tolerance

At high temperatures, common events occur at the cell level, *i.e.*, the generation and reaction of reactive oxygen species (ROSs). Excessive ROS production above the threshold is undoubtedly harmful to all cell compounds because it hurts cell metabolism. To curb the detrimental effect of ROS, a complex antioxidative defense system in plants is developed, namely superoxide dismutase (SOD), catalase, guaiacol peroxidase, ascorbate peroxidase, dehydroascorbate reductase, glutathione reductase, and glutathione S-transferase and non-enzymatic antioxidants for example flavonoids, anthocyanin, carotenoids and ascorbic acid [41]. O_2- is converted to H_2O_2 by superoxide dismutase, while the dismutation of H_2O_2 is carried out by catalase and peroxidase. Catalase destroys H_2O_2 by separating it from H_2O and O_2, but peroxidase requires a reduction equivalent to the elimination of H_2O_2. Guaiacol peroxidase, as a donor of electrons, requires a phenol compound Guaiacol peroxidase to decompose H_2O_2, while Ascorbate peroxidase utilizes a reduced form of ascorbate (AsA) to protect the cell against the injurious effect of H_2O_2. The most minacious forms of reactive oxygen species, such as OH^-, O_2^-, and H_2O_2, are scavenged by a reduced form of ascorbate (AsA) through the action of ascorbate peroxidase. Activation of different antioxidant enzymes occurs at different temperature ranges, as the activation is temperature-sensitive. Catalase, Ascorbate peroxidase, and superoxide dismutase activity increased up to 50 °C but declined at 50 °C, while Peroxidase and glutathione reductase activity declined at all temperatures ranging from 20°C to 50 °C [42]. Total antioxidant activity is maximum at 35 °C to 40 °C in tolerant varieties and 30 °C for susceptible ones. One of the major antioxidant defense systems against ROS is the super oxidase dismutase that converts superoxide into H_2O_2, while catalase (CAT), guaiacol peroxidase (GPX) and ascorbate peroxidase (APX) detoxify H_2O_2.

4.3. The Function of Heat Stress Proteins (Hsps) in Heat Tolerance

Hsp is also known as stress-induced protein or stress protein [43]. Almost all stress results in the expression of genes and the synthesis of stress proteins. The stress agent causes immediate blockage of important metabolic processes such as DNA replication, transcription, export of mRNA, and translation until cells are recovered. There are five molecular-mass Hsps: Hsp100, Hsp90, Hsp70, Hsp60, and small heat shock proteins (sHsp). Normally, pants have 20 sHsps, and there are 40 kinds of sHsps in a plant species. Variations in these stress proteins indicate sensitivity to heat stress. Stress proteins are responsible for the molecular chaperones that regulate protein folding and accumulation. Chaperon proteins are heat shock proteins that can prevent non-specific aggregation by combining with non-original proteins and precipitating the protein refolding during heat stress.

4.4. The Function of Exogenous Phyto-protectants in Heat Tolerance

The plants have the adaptative mechanism to accumulate proline, glycine betaine, and trehalose to prevent the injurious effect of abiotic stresses, especially heat stress. There is a lack of the accumulation of these substances in thermo-sensitive plants. In those plants, heat tolerance can be obtained by the exogenous application of the osmoprotectants. The application of these osmoprotectants reduces H_2O_2 production and protects the growing tissues from heat stress. Similarly, the exogenous application of phytohormones helps in adaptation to heat stress, like ABA, and lessens the heat stress manifestations by increasing the activities of super oxidase dismutase, catalase and peroxidase, which reduces the H_2O_2 production. Indole Acetic Acid changes the plant architecture in response to heat stress by regulating root growth and gray stimulation. Under high-temperature stress, the signals induced by IAA change due to increased IAA level that switches the developmental progress. Thus, IAA helps in adaptation under HT stress. Both growth promoters and growth preserver hormones help overcome heat stress. A signaling molecule like nitric acid, depending on its concentration, has both beneficial and harmful impacts on plant tissues. Nitric acid, due to its antioxidant property, acts as a signal for an antioxidant defense system that scavenges the reactive oxygen species under high-temperature stress. In many plant studies, trace elements have shown an advantageous effect on heat-stressed plants. Si treatment on plants reduces the HT stress by preventing electrolyte leakage and by maintaining the membrane integrity of plant cells. Also, some studies proved that foliar spray of selenium (Se) improves the rate of photosynthesis, stomatal transmittance, and rate of transpiration by 13.2%, 12.4%, and 8.11%, respectively.

Osmo-protectants	•Proline •Betaine •Trehalose

Phyto-hormones	•Abscisic acid •Auxins •Gibberellins •Jasmonic Acid •Salicylic Acid •Brassinosteriods

Signaling molecules	•Nitric oxide

Trace elements	• Selenium • Silicon

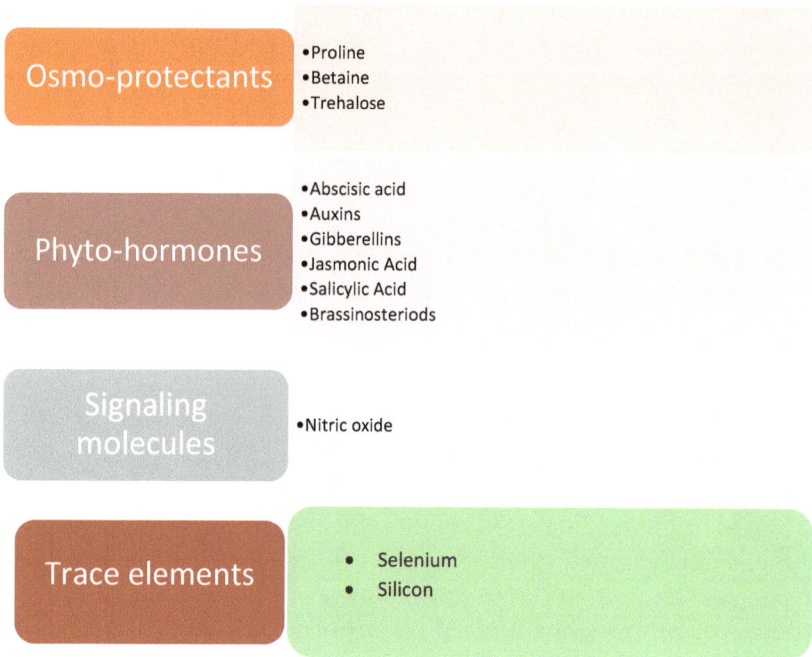

4.5. Genetic Engineering Approach For Heat Tolerance

The genetically engineered approach has led to the development of plants that can tolerate heat stress. This approach involves the insertion of the gene of interest in the recipient genotypes to develop the heat tolerance mechanism. The use of molecular markers and gene transformation also helps in the development of HT-tolerant plants. The plasticity of plants' genomes, such as directed mutations and epigenetics, such as methylation, chromatin remodeling, and histone acetylation changes, enable long-term adaptation to biological stress and are necessary for the long-term survival of plant genotypes. The use of molecular markers and genetic engineering approaches has decreased the loss due to HT. Quantitative trait loci (QTLs) mapping has also led to genetic mechanisms to tolerate various abiotic stresses.

5. PLANT RESPONSE TO LOW-TEMPERATURE STRESS: AN OVERVIEW

Low-temperature stress is an important environmental factor affecting plant growth and productivity when temperatures are lower than optimal temperatures, leading to significant crop losses.

The chilling temperature is the low temperature that causes injury but does not form ice crystals in the plant tissues. While, freezing temperature leads to the formation of ice crystals within the plant tissues. Both chilling and freezing stress cause water deficit conditions in the plant cells, which are called cold stress. Plants grown under cold (temperate) climates are deemed to be chilling tolerant while freezing tolerance can be achieved by the process of cold acclimation, *i.e.*, exposure to cold but non-freezing temperatures. Plants that are grown under tropical and sub-tropic conditions lack the mechanism of cold acclimation, and it may affect crop survival, photosynthesis, water movement, and, finally, yield.

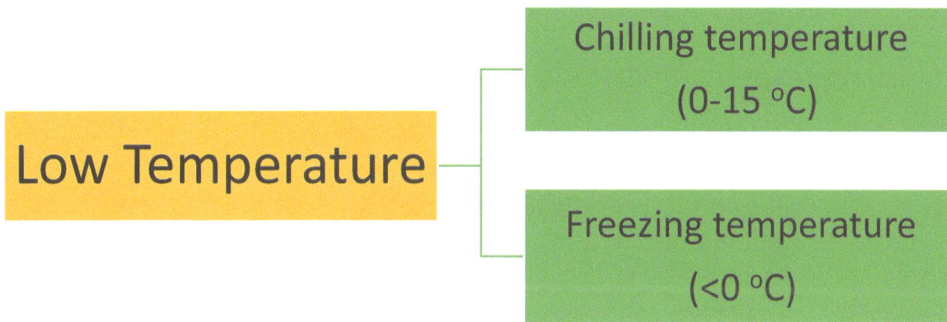

| Low Temperature | Chilling temperature (0-15 °C) |
| | Freezing temperature (<0 °C) |

5.1. Chilling Injury

CHILLING INJURY
• Take place at a low but non-freezing temperature
• Occurs in tropical and sub-tropical plants at 10 °C to 25 °C
• In temperate plants at 0 °C to 15 °C
• Indicated by cytological and physiological changes

The term 'Chilling injury' was given by German plant physiologist 'Molish' in 1897.

5.2. Cytological Changes Caused by Chilling Injury

The detrimental effect of chilling injury is;

- Destruction of cell membranes leads to deprivation of cell compartmentation.
- Swelling and bursting of the plasmalemma.
- The devastation of the endoplasmic reticulum and vesiculation of its membranes.
- Change in Golgi bodies.
- Visible changes in the structure of mitochondria;

1. Their swelling and degeneration.

2. Matrix enlightenment.

3. Cristae shortening and decrease in their number, leading to a reduction of oxidative phosphorylation.

- Destruction of the chloroplast membranes.
- Accumulation of lipid bodies and disappearance of starch grains.
- Causes accelerated cell differentiation but reduced cell growth.
- Cytoplasmic viscosity decreases due to changes in the colloid-chemical properties of cytoplasm.
- Long-term chilling causes the coagulation of structural proteins.
- Ribosomes completely disappear.

5.3. Physiological Changes Caused by Chilling Injury

Chilling injury disrupts various physiological processes;

| Water Regimes | Mineral nutrition | Photosynthesis | Respiration rate |

5.4. Water Regimes

The cooling affects all components of the water system, resulting in water loss, leading to strong wilting. It is based on two main elements;

- The capability of the root system to absorb water and move it into the shoot system decreases and also reduces the capacity to close the stomata under water deficit conditions.

- The rapid drop in water potential in leaves is due to an inadequate water supply. The chilling damage can be prevented by maintaining the water regime.

5.5. Mineral Nutrition

Chilling has a detrimental effect on plant mineral nutrition;

- The absorption and movement of ions by roots are affected.
- The decrease in the nutrient content of the plant is due to the uneven distribution of nutrients within the plant organs.
- Reduction in the activity of nitrogen reductase, decrease in nitrogen assimilation in proteins and amino acids.
- Increase in inorganic phosphorus content and decrease in organic phosphorus.
- Decline in oxidative phosphorylation and change the membrane integrity.
- Lowers the supply of ATP to H ion transporting ATPase and decreases the ion permeability coefficients.

5.6. Respiration Rate

At the coldest temperature, the breathing rate changes. There is evidence that the breathing rate under cold temperatures is declining and increasing. This decrease is caused by the destruction of mitochondrial structures, a decrease in motion energy, and the inhibition of enzymes. The increase in respiration rate is believed to be the result of the uncoupling of oxidative phosphorylation.

5.7. Photosynthesis Rate

Photosynthesis inhibition occurs due to various physiological causes;

- Hampered transport of glucose from leaves to the phloem.
- Stomata limitation.
- The devastation of photosynthetic components.
- Injury in the water-splitting complex of PSI.
- Impeded electron transport.
- The inactivation and inhibition of the synthesis of enzymes from the C4 pathway and the Calvin cycle.
- Photo oxidative damage to photosystems in the chloroplast membranes.

6. MECHANISM FOR CHILLING TOLERANCE

Various techniques to tolerate chilling temperature are grouped under three categories;

Thermal Effect	Chemical treatment	Cellular and genetic engineering
• Low temperature hardening • Thermal conditioning • Intermediate warming • Effect of Heat stress	• Effect of trace elements • Synthetic growth regulators • Antioxidants	• Gene tranfer • Selective markers

6.1. Thermal Effect

- Low-temperature hardness is a process associated with the protein synthesis system, which is accompanied by the restructuring of the plant hormone system.
- Thermal conditioning or preconditioning is the process linked with the change in plant response to chilling before exposure to reduced temperature. It leads to acclimation, which develops resistance in plants against chilling.
- Intermediate warming is the process of a temporary exposure of chilled plants to heat that allows the chilled tissues to regenerate the compounds in plant tissues that were destroyed during chilling.
- Exposure to heat stress induces chilling resistance in plants due to the development of new mRNAs and proteins.

6.2. Chemical Treatment

- Synthetic phytohormones and growth regulators are the most promising compounds that increase the chilling resistance in chilling sensitive plants, such as cytokinin and ABA are the most effective phytohormones. Nonhormonal growth regulators, namely paclobutrazol, chlorophyllin chloride, and other triazoles, improve the chilling resistance.
- Antioxidants and free radicle treatment with sodium benzoate, ascorbate, tocopherol, and propyl gallate help in preventing the degradation of unsaturated fatty acids and declines the chilling damage to plants.

6.3. Cellular and Genetic Engineering

- It allows fundamental changes in the resistance of chilling sensitive plants.
- It involves large genetic variations in components and control sensitivity but is based on genetic transfer technologies, transformations and selection markers.

6.4. Freezing Injury

FREEZING INJURY
• Occurs in plants when the temperature is below 0°C.
• As the temperature of the cells decreases and the space between the cells supercool
• The formation of extracellular ice is due to the presence of the concentration of the solvent.
• The plasma membrane prevents intracellular hydration.
• Because there is no contact between the ice and intercellular water, a vapor pressure gradient is developed from inside to outside.
• The equilibrium is achieved by evaporating water from the cell and creating extracellular ice or intracellular ice.
• The injury occurs only upon thawing.

Sources of freezing injury in plants

Freezing of soil water	Freezing of fluid within plant
• The water present in the pore spaces between soil particles is available to plants. it freezes at -2°C, making soil water unavailable to plants.	• It's a major concern since it disrupts the structure and function of cells and tissues. It is largely caused by the production of ice crystals.

Types of freezing that occur in plant cells and tissues;

Vitrification	• Solidifaction of the cellular content into amorphous state. It occurs due to rapid freezing at low temperature. (decrease 30 °C/min).
Crystallization/ ice formation	• Crystallization of ice either extracellularly or intracellularly by graduale drop in temperature

7. MECHANISM FOR FREEZING TOLERANCE

7.1. Adaptation

Plants that live in arctic and temperate conditions where the temperature is likely to fall, must have the potential to deal with internal freezing and to survive under extended drought caused by soil freezing. Both problems are solved by small flowering plants by devoting the winter to seeds and renewing the vegetative growth annually. Seeds elude the freezing temperature by stowing in the ground under-insulated snow cover.

7.2. Avoidance

Avoid ice crystal formation within the plant by existing in a supercooled state. However, this is not a winter strategy for small plants because they are very cold under freezing at about 4 to 5 degrees Celsius. In the case of woody plants, cells of xylem tissues remain super cool throughout the winter even though extracellular ice formation occurs in other tissues. Ice crystal formation in the xylem is prevented by waterproof barriers.

7.3. Tolerance

• Plant hardening occurs as winter approaches to tolerate freezing. The plant undergoes necessary changes for survival under ice growth within tissues.

- The plant undergoes various alterations before the temperature gets dropped. The changes that occur are; an increase in cytoplasmic solute concentration to act as also to protect the freezing concentration of other solvents that may cause damage at higher concentrations while buffering the freezing concentrations of other solvents, also an alteration in the membrane that increases the fluidity of membranes at freezing temperature, and maintains membrane integrity to low water levels.

CONCLUSION

Both low- and high-temperature stress has a major negative impact on the production of crops all over the world by reducing their growth and development and, eventually, reproduction. Under climate change, the greenhouse effect is a major contributor to the rise of global temperatures. Plants undergo various morphological and physiological changes that cause severe cellular damage or even cell death that leads to the lethal collapse of the cellular organization when faced with high-temperature stress. Fortunately, plants have developed various adaptation strategies, namely avoidance and tolerance mechanisms to tolerant high-temperature stress. Theoretically, chilling injury can lead to the disruption of all physiological processes in plants, including water regimes, photosynthesis rate, respiration rate, and metabolism. To avoid chilling injury, various mechanisms are adapted, such as thermal effects, chemical treatments, and cellular and genetic engineering techniques. Freezing injury causes the ice crystal formation intracellularly or extracellularly, leading to membrane instability and leakage of ions. Protecting plants from freezing injury can be achieved through the lowering of the freezing point by solutes, and additionally, some plants allow deep supercooling or plant hardening to survive under ice conditions.

REFERENCES

[1]   Watanabe T, Kume T. A general adaptation strategy for climate change impacts on paddy cultivation: special reference to the Japanese context. Paddy Water Environ 2009; 7(4): 313-20.
        [http://dx.doi.org/10.1007/s10333-009-0179-5]

[2]   Hasanuzzaman M, Hossain MA, da Silva JA, Fujita M. Plant response and tolerance to abiotic oxidative stress: antioxidant defense is a key factor InCrop stress and its management: perspectives and strategies. Dordrecht: Springer 2012; pp. 261-315.

[3]   Wigley TML, Raper SCB. Interpretation of high projections for global-mean warming. Science 2001; 293(5529): 451-4.
        [http://dx.doi.org/10.1126/science.1061604] [PMID: 11463906]

[4]   Wahid A, Gelani S, Ashraf M, Foolad M. Heat tolerance in plants: An overview. Environ Exp Bot 2007; 61(3): 199-223.
        [http://dx.doi.org/10.1016/j.envexpbot.2007.05.011]

[5]   Hasanuzzaman M, Nahar K, Alam M, Roychowdhury R, Fujita M. Physiological, biochemical, and molecular mechanisms of heat stress tolerance in plants. Int J Mol Sci 2013; 14(5): 9643-84.
        [http://dx.doi.org/10.3390/ijms14059643] [PMID: 23644891]

[6]   Mittler R, Blumwald E. Genetic engineering for modern agriculture: challenges and perspectives. Annu Rev Plant Biol 2010; 61(1): 443-62.
[http://dx.doi.org/10.1146/annurev-arplant-042809-112116] [PMID: 20192746]

[7]   Lobell DB, Schlenker W, Costa-Roberts J. Climate trends and global crop production since 1980. Science 2011; 333(6042): 616-20.
[http://dx.doi.org/10.1126/science.1204531] [PMID: 21551030]

[8]   McClung CR, Davis SJ. Ambient thermometers in plants: from physiological outputs towards mechanisms of thermal sensing. Curr Biol 2010; 20(24): R1086-92.
[http://dx.doi.org/10.1016/j.cub.2010.10.035] [PMID: 21172632]

[9]   Ahuja I, de Vos RCH, Bones AM, Hall RD. Plant molecular stress responses face climate change. Trends Plant Sci 2010; 15(12): 664-74.
[http://dx.doi.org/10.1016/j.tplants.2010.08.002] [PMID: 20846898]

[10]   Kumar S, Kaur R, Kaur N, *et al.* Heat-stress induced inhibition in growth and chlorosis in mungbean (Phaseolus aureus Roxb.) is partly mitigated by ascorbic acid application and is related to reduction in oxidative stress. Acta Physiol Plant 2011; 33(6): 2091-101.
[http://dx.doi.org/10.1007/s11738-011-0748-2]

[11]   Johkan M, Oda M, Maruo T, Shinohara Y. Crop production and global warming. Global warming impacts-case studies on the economy, human health, and on urban and natural environments. 5: 139-52.2011;

[12]   Piramila BH, Prabha AL, Nandagopalan V, Stanley AL. Effect of heat treatment on germination, seedling growth and some biochemical parameters of dry seeds of black gram. Int J Pharm Phytopharmacol Res 2012; 1: 194-202.
[PMID: 86300926]

[13]   Yang X, Liang Z, Lu C. Genetic engineering of the biosynthesis of glycinebetaine enhances photosynthesis against high temperature stress in transgenic tobacco plants. Plant Physiol 2005; 138(4): 2299-309.
[http://dx.doi.org/10.1104/pp.105.063164] [PMID: 16024688]

[14]   Wang JZ, Cui LJ, Wang Y, Li JL. Growth, lipid peroxidation and photosynthesis in two tall fescue cultivars differing in heat tolerance. Biol Plant 2009; 53(2): 237-42.
[http://dx.doi.org/10.1007/s10535-009-0045-8]

[15]   Sato S, Kamiyama M, Iwata T, Makita N, Furukawa H, Ikeda H. Moderate increase of mean daily temperature adversely affects fruit set of Lycopersicon esculentum by disrupting specific physiological processes in male reproductive development. Ann Bot (Lond) 2006; 97(5): 731-8.
[http://dx.doi.org/10.1093/aob/mcl037] [PMID: 16497700]

[16]   Maheswari M, Yadav SK, Shanker AK, Kumar MA, Venkateswarlu B. Overview of plant stresses: Mechanisms, adaptations and research pursuit InCrop stress and its management: Perspectives and strategies. Dordrecht: Springer 2012; pp. 1-18.

[17]   Foolad MR. Abiotic Stresses: Plant Resistance Through Breeding and Molecular Approaches.

[18]   Yun-Ying CA, Hua DU, Li-Nian YA, Zhi-Qing WA, Shao-Chuan ZH, Jian-Chang YA. Effect of heat stress during meiosis on grain yield of rice cultivars differing in heat tolerance and its physiological mechanism. Zuo Wu Xue Bao 2008; 34(12): 2134-42.

[19]   Hurkman WJ, Vensel WH, Tanaka CK, Whitehand L, Altenbach SB. Effect of high temperature on albumin and globulin accumulation in the endosperm proteome of the developing wheat grain. J Cereal Sci 2009; 49(1): 12-23.
[http://dx.doi.org/10.1016/j.jcs.2008.06.014]

[20]   Ahamed KU, Nahar K, Fujita M, Hasanuzzaman M. Variation in plant growth, tiller dynamics and yield components of wheat (*Triticum aestivum* L.) due to high temperature stress. Adv Agric Bot 2010; 2(3): 213-24.

[21]   Suwa R, Hakata H, Hara H, *et al.* High temperature effects on photosynthate partitioning and sugar metabolism during ear expansion in maize (*Zea mays* L.) genotypes. Plant Physiol Biochem 2010; 48(2-3): 124-30.
[http://dx.doi.org/10.1016/j.plaphy.2009.12.010] [PMID: 20106675]

[22]   Tubiello FN, Soussana JF, Howden SM. Crop and pasture response to climate change. Proc Natl Acad Sci USA 2007; 104(50): 19686-90.
[http://dx.doi.org/10.1073/pnas.0701728104] [PMID: 18077401]

[23]   Crawford AJ, McLachlan DH, Hetherington AM, Franklin KA. High temperature exposure increases plant cooling capacity. Curr Biol 2012; 22(10): R396-7.
[http://dx.doi.org/10.1016/j.cub.2012.03.044] [PMID: 22625853]

[24]   Cheng L, Zou Y, Ding S, *et al.* Polyamine accumulation in transgenic tomato enhances the tolerance to high temperature stress. J Integr Plant Biol 2009; 51(5): 489-99.
[http://dx.doi.org/10.1111/j.1744-7909.2009.00816.x] [PMID: 19508360]

[25]   Kolb PF, Robberecht R. High temperature and drought stress effects on survival of Pinus ponderosa seedlings. Tree Physiol 1996; 16(8): 665-72.
[http://dx.doi.org/10.1093/treephys/16.8.665] [PMID: 14871688]

[26]   Morales D, Rodríguez P, Dell'Amico J, Nicolás E, Torrecillas A, Sánchez-Blanco MJ. High-temperature preconditioning and thermal shock imposition affects water relations, gas exchange and root hydraulic conductivity in tomato. Biol Plant 2003; 46(2): 203-8.
[http://dx.doi.org/10.1023/B:BIOP.0000022252.70836.fc]

[27]   Mazorra LM, Núñez M, Hechavarria M, Coll F, Sánchez-Blanco MJ. Influence of brassinosteroids on antioxidant enzymes activity in tomato under different temperatures. Biol Plant 2002; 45(4): 593-6.
[http://dx.doi.org/10.1023/A:1022390917656]

[28]   Taiz L, Zeiger E. (eds) Plant physiology, 5th edn. Sinauer Associates, Sunderland 671-81.2006;

[29]   Wahid A, Close TJ. Expression of dehydrins under heat stress and their relationship with water relations of sugarcane leaves. Biol Plant 2007; 51(1): 104-9.
[http://dx.doi.org/10.1007/s10535-007-0021-0]

[30]   Shah NH, Paulsen GM. Interaction of drought and high temperature on photosynthesis and grain-filling of wheat. Plant Soil 2003; 257(1): 219-26.
[http://dx.doi.org/10.1023/A:1026237816578]

[31]   Young LW, Wilen RW, Bonham-Smith PC. High temperature stress of Brassica napus during flowering reduces micro- and megagametophyte fertility, induces fruit abortion, and disrupts seed production. J Exp Bot 2004; 55(396): 485-95.
[http://dx.doi.org/10.1093/jxb/erh038] [PMID: 14739270]

[32]   Bañon S, Fernandez JA, Franco JA, Torrecillas A, Alarcón JJ, Sánchez-Blanco MJ. Effects of water stress and night temperature preconditioning on water relations and morphological and anatomical changes of Lotus creticus plants. Sci Hortic (Amsterdam) 2004; 101(3): 333-42.
[http://dx.doi.org/10.1016/j.scienta.2003.11.007]

[33]   Morales D, Rodríguez P, Dell'Amico J, Nicolás E, Torrecillas A, Sánchez-Blanco MJ. High-temperature preconditioning and thermal shock imposition affects water relations, gas exchange and root hydraulic conductivity in tomato. Biol Plant 2003; 46(2): 203-8.
[http://dx.doi.org/10.1023/B:BIOP.0000022252.70836.fc]

[34]   Omae H, Kumar A, Kashiwaba K, Shono M. Adaptation to high temperature and water deficit in the common bean (Phaseolus vulgaris L.) during the reproductive period. J Bot 2012; 2012(2).
[http://dx.doi.org/10.1155/2012/803413]

[35]   Berry JA, Raison JK. Responses of macrophytes to temperature In Physiological plant ecology I. Berlin, Heidelberg: Springer 1981; pp. 277-338.

[36]   Steponkus PL. Responses to extreme temperatures Cellular and sub-cellular bases InPhysiological Plant Ecology I. Berlin, Heidelberg: Springer 1981; pp. 371-402.

[37]   Asada K. Production and scavenging of reactive oxygen species in chloroplasts and their functions. Plant Physiol 2006; 141(2): 391-6.
[http://dx.doi.org/10.1104/pp.106.082040] [PMID: 16760493]

[38]   Møller IM, Jensen PE, Hansson A. Oxidative modifications to cellular components in plants. Annu Rev Plant Biol 2007; 58(1): 459-81.
[http://dx.doi.org/10.1146/annurev.arplant.58.032806.103946] [PMID: 17288534]

[39]   Garg N, Manchanda G. ROS generation in plants: Boon or bane? Plant Biosyst 2009; 143(1): 81-96.
[http://dx.doi.org/10.1080/11263500802633626]

[40]   Montillet JL, Chamnongpol S, Rustérucci C, *et al.* Fatty acid hydroperoxides and H_2O_2 in the execution of hypersensitive cell death in tobacco leaves. Plant Physiol 2005; 138(3): 1516-26.
[http://dx.doi.org/10.1104/pp.105.059907] [PMID: 15980200]

[41]   Suzuki N, Miller G, Morales J, Shulaev V, Torres MA, Mittler R. Respiratory burst oxidases: the engines of ROS signaling. Curr Opin Plant Biol 2011; 14(6): 691-9.
[http://dx.doi.org/10.1016/j.pbi.2011.07.014] [PMID: 21862390]

[42]   Chakraborty U, Pradhan D. High temperature-induced oxidative stress in *Lens culinaris*, role of antioxidants and amelioration of stress by chemical pre-treatments. J Plant Interact 2011; 6(1): 43-52.
[http://dx.doi.org/10.1080/17429145.2010.513484]

[43]   Gupta SC, Sharma A, Mishra M, Mishra RK, Chowdhuri DK. Heat shock proteins in toxicology: How close and how far? Life Sci 2010; 86(11-12): 377-84.
[http://dx.doi.org/10.1016/j.lfs.2009.12.015] [PMID: 20060844]

CHAPTER 6

Molecular Studies and Metabolic Engineering of Phytohormones for Abiotic Stress Tolerance

Sekhar Tiwari[1] and **Ravi Rajwanshi**[2,*]

[1] *School of Sciences, P P Savani University, Surat, Gujarat, India*

[2] *Discipline of Life Sciences, School of Sciences, Indira Gandhi National Open University, New Delhi, India*

Abstract: Agricultural productivity across the world is affected by varied abiotic stresses, which require the development of crops tolerant to unfavorable conditions without considerable yield loss. In recent times, considerable importance has been given to phytohormones because of their versatile functions in plant responses to environmental constraints and for their role in the regulation and coordination of the growth and development of plants. Research on phytohormones has shed light on the role of classical and new members of phytohormones in alleviating the harmful effects of abiotic stresses on crop plants, so understanding phytohormone metabolism and its engineering could be a potent and novel approach for developing climate-resilient crops. The present chapter presents a short description of classical and new members of phytohormones and their role in alleviating varied abiotic stresses. Furthermore, molecular and genetic engineering efforts undertaken for the development of crops tolerant to abiotic stresses are also presented along with research gaps and challenges for the utilization of phytohormones for the development of abiotic stress-tolerant plants.

Keywords: Auxins, Brassinosteroids, Phytohormones, Phytohormone engineering, Salicylic acid, Strigolactones.

1. INTRODUCTION

The rising human population, coupled with climate changes, is putting unprecedented pressure on the agricultural system, necessitating a significant rise in the production and yield of crops. Several types of abiotic and biotic stresses are the key reasons for low agricultural productivity [1]. Abiotic stresses like drought, salinity, temperature (chilling or freezing), heat, ultraviolet radiations, weak and intense light, and gaseous pollutants of the atmosphere (sulfur dioxide, ozone) are affecting crop productivity to a great extent [2]. By the end of 2050,

---

* **Corresponding author Ravi Rajwanshi:** Discipline of Life Sciences, School of Sciences, Indira Gandhi National Open University, New Delhi, India; E-mail: rrajwanshi@gmail.com

Jen-Tsung Chen (Ed.)

agricultural production would have to increase by 70% to feed an additional nearly two and a half billion people [3]. All these factors have necessitated the development of stress-tolerant cultivars to mitigate a wide array of stresses in different agro-climatic conditions. Plants have distinct and much more complicated systems for environmental stress response and tolerance than animals [4]. Owing to complexity, traditional breeding strategies have had limited success with stress tolerance traits. The development of plant biotechnological tools and the use of genetic engineering has offered an effective solution to address the limitation of conventional plant breeding for the development of plants tolerant to stresses [5]. In the recent past, because of multidimensional functions in response to abiotic stress, phytohormones have received great consideration.

Phytohormones are substances that are produced in very small amounts yet have the ability to control a range of cellular processes in plants. They function as signaling molecules in plant species to convey cellular activity [6]. Phytohormones serve an important function in the regulation of numerous signal transduction pathways at the time of perceived stress. They control both exterior and internal stimuli [7]. Besides classical phytohormones, newer members of phytohormones may offer effective targets for metabolic engineering for developing climate-resilient crops with a higher yield. The present chapter reviewed phytohormones and their functions in various aspects of abiotic stress tolerance, along with their molecular and metabolic engineering efforts undertaken to develop plants to boost agricultural production and productivity.

2. PHYTOHORMONES MEDIATED ABIOTIC STRESS TOLERANCE

To respond to diverse external and internal stimuli, plants must control their development and growth [8]. These responses are mediated by phytohormones, a varied group of signaling chemicals present in minute amounts in cells. Their importance in enabling plant acclimatization to ever-changing environments through nutrient allocation, source/sink transitions, and regulating development and growth has long been recognized [9]. Plant response to abiotic stressors is influenced by a variety of factors, but phytohormones, the most significant endogenous chemicals for influencing molecular and physiological responses, offer adaptive responses for the plant in stressed conditions [9]. Phytohormones can operate at their production location or anywhere in the plant once transported [10]. Plant development and plastic growth rely heavily on phytohormones.

2.1. Abscisic Acid (ABA)

It is an abiotic stress hormone because of its particular and responsive effect on plant adjustment or adaptation to abiotic stresses. It gets its name from its function related to the abscission of plant leaves. The plastidial methylerythritol 4-phosphate (MEP) pathway produces it as an isoprenoid plant hormone. The creation of storage lipids and proteins, embryo morphogenesis, stomatal opening, seed development and dormancy are only a few of the physiological processes and developmental stages in which ABA plays a role [11]. The function of ABA during stress resilience has attracted a lot of interest since it is regarded as an important messenger during adaptive responses of plants to abiotic stress. The rapid increment of endogenous abscisic acid indicates that the plant is responding to environmental stresses by activating certain signaling pathways and altering gene expression levels [12]. ABA upon transcription influences up to 10% of protein-encoding DNA sequences, according to Nemhauser *et al.* [13]. Abscisic acid also serves as an internal stimulus for plants, allowing them to thrive in harsh environments [14]. Under water-stressed situations, ABA is critical for plants to communicate with their shoots about stressful issues near the roots, which leads to water-saving anti-transpirant behavior such as reduced leaf growth and stomatal closure [15]. Under drought stress [16] and nitrogen deprivation [17], abscisic acid is also implicated in vigorous root development and other structural alterations. ABA is involved in the manufacture of protective proteins such as dehydrins and LEA proteins, as well as the translation of various genes that are involved in the stress response [18, 19]. Desiccation tolerance is conferred through the upregulation of mechanisms engaged in the manufacture of antioxidant and osmoprotectant enzymes and cell turgor pressure maintenance by ABA [20]. Zhang *et al.* [21] observed a proportionate rise in abscisic acid concentration in plants that are exposed to salinity.

2.2. Auxins (IAA)

Even though IAA has been researched for almost a century, the production, transport, and signaling routes remain unknown [22]. However, some interconnected routes for auxin production in plants have been proposed so far, including one Trp-independent and four Trp-dependent pathways [23]. Being a multifunctional phytohormone, Indole-3-acetic acid is important for plant development and growth as well as in regulating and coordinating plant development and growth in stressful conditions [24]. The availability of an auxin production, signaling, and transportation system in single-celled green algae has demonstrated auxin's evolutionary relevance in plant adaptation to various terrestrial conditions [25]. Despite recent gains in our knowledge of auxin's role in plant development and growth, its function as a stress response regulator is still unknown [24]. Surprisingly, there is mounting proof that Indole-3-acetic acid has a key role in plant salinity stress response [9, 26]. It promotes shoot and root growth in plants that are exposed to salt or heavy metals [27, 28]. Auxin induces the transcription process of the many numbers of protein-coding genes known as primary auxin response genes, which have been detected and characterized in plants such as soybean, arabidopsis and rice [29]. Auxin is thought to be an important component of defensive responses because it regulates several genes and mediates the interaction between biotic and abiotic stress responses [30]. The discovery of new genes implicated in response to stress, on the other hand, might have shown to be a crucial target for conferring tolerance to major crops against abiotic stresses [5].

2.3. Cytokinins (CKs)

Cytokinins are master regulators throughout plant development and growth since they have a function in varied plants' growth and developmental processes [31, 32]. Abiotic stress [12], such as salinity [32] and drought [31], is indicated by fluctuation in endogenous cytokinin levels in response to stress conditions. Changes in the activity of cytokinin perception machinery or metabolic enzymes in mutants and transgenic cells/tissues suggest their critical role in diverse agricultural attributes, including stress tolerance as well as productivity [33]. However, plant response to cytokinins has primarily been studied in the context of their external application, stressful situations that have been shown to increase endogenous levels of CKs through absorption and increased production [34]. They also help to release seeds from the dormant stage, in contrast to abscisic acid's restriction of seed germination [30]. CKs are frequently referred to as ABA antagonists [35]. Reduced CK content and ABA accumulation cause a rise in the ABA/CK ratio in water-stressed plants. Decreased cytokinin levels promote apical dominance, which, in combination with abscisic acid-controlled stomatal aperture modulation, promotes drought stress adaption [12].

2.4. Ethylene (ET)

It is a gaseous plant hormone that regulates stress responses and is engaged in various stages of plant development and growth, including floral senescence, petal and leaf abscission and fruit ripening [36, 37]. Cyclic non-protein amino acid ACC and S-adenosyl-L-methionine (AdoMet) are used to biosynthesize it from methionine. ACC synthase catalyzes the conversion of AdoMet to ACC, while ACC oxidase enzyme catalyzes the reformation of ACC into ethylene by ACC oxidase. Plants' endogenous ET levels are affected by abiotic stressors such as salinity and low temperature. Higher ET concentrations were found to result in improved tolerance [38]. ET is also important in plants' defense response to heat stress [39]. Environmental stress responsible for ET buildup increases the chances of plants thriving in these extreme environments [36]. ET is thought to work by modulating gene expression, which is thought to be one of the ethylene signal's effectors [40]. When ET is combined with other phytohormones like SA and JA, it can have a cooperative effect. These are the most important players in controlling plant defense against diseases and pests [7]. The production, transport, and storage of these hormones activate a series of plant defense signaling pathways [41]. ABA and ET appear to function antagonistically or synergistically to influence plant development and growth, according to Yin *et al.* [42].

2.5. Gibberellins (GAs)

They belong to a wide family of tetracyclic diterpenoid carboxylic acids. Among them, only a few members play a role in higher plants as growth hormones, with GA4 and GA1 being the most frequent [43]. Fruit and flower development, trichrome and flower initiation, stem elongation, leaf expansion, and seed germination are all improved by GAs [44]. Plants require them for growth-stimulatory actions throughout their life cycle. Transitions between developmental phases are also encouraged [45]. Surprisingly, proof of their critical involvement in response to abiotic stress and adaptability is growing [45]. Using seedlings of *Arabidopsis thaliana*, the experiment was conducted to find out the function of GAs in response to osmotic stress [46, 47]. In a variety of developmental and stimulus-response processes, GAs also interfere with all of the other phytohormones [48]. Depending on the signaling context and tissue, ET and GA interactions involve both positive and negative mutual regulation [48].

2.6. Brassinosteroids (BRs)

They are a relatively recent class of polyhydroxy steroidal phytohormones with remarkable development and growth-promoting properties. They were discovered and identified in the pollen of the *Brassica napus*; this plant is also called the rape plant. Seventy different BRs are yielded by the plants. The 3 most bioactive BRs, brassinolide, 28-homobrassinolide, and 24-epibrassinolide are widely employed in experimental and physiological research [49]. Roots, shoots, leaves, vascular cambium, seeds, fruits, flower buds, and pollen are just a few of the places where they can be found [50]. They are engaged in a wide range of developmental activities, such as fruit and flower development, floral initiation and root and stem growth [50]. Recent discoveries, on the other hand, reveal that BRs and related chemicals may play a stress-impact-reducing role in a variety of plants exposed to various abiotic pressures like organic contaminants [51], metals/metalloids [52], flooding [53], drought [54], light [55], soil salinity [56], cold [57], and high temperature [58] are examples of abiotic stresses. Vardhini and Anjum [2] reviewed recent studies that showed enormous promise for BRs and related substances in modulating antioxidant defense system components in reaction to and counteracting oxidative bursts caused by abiotic stress. However, there is a lot of need for more research into their biosynthesis sites, routes, enzymology, interactions with microbes, animals and fungi, and the realization of their strong uses in stress and developmental physiology and source-sink relationships [9].

2.7. Jasmonates (JAs)

They are cyclopentanone phytohormones generated from the metabolism of jasmonic acid (JA) and methyl jasmonate (MeJA). They are part of the membrane protein, which is found throughout the plant kingdom. Jasmonic acid is the free acid of jasmonate. These multifunctional chemicals play a role in reproductive activities, indirect and direct defensive responses, secondary metabolism, senescence, flowering, and fruiting, all of which are important for plant development and survival [59, 60]. The most common well-characterized and well-known of the JAs is JA. JA triggers the plant's defensive system against pathogenic assault as well as external challenges such as low temperature, salt, and drought, in addition to developmental roles [61, 62]. Environmental stressors such as salt [63], UV irradiation [63], and drought [64, 65] cause JAs, which are important signaling molecules. They have a lot of potential for reducing a variety of dangerous environmental pressures [66]. Exogenous treatment of MeJA to soybean seedlings successfully reduced salt stress symptoms [67]. Under salt stress, endogenous levels of JA were raised in rice roots, which was found to mitigate the negative effects of high salt stress [68]. By activating the antioxidant system, heavy metal-mediated stress is alleviated by JAs [69]. Through the build-up of phytochelatins, MeJA provides resilience to Cu and Cd stress in *A. thaliana* [70].

2.8. Salicylic Acid (SA)

It is a phenolic molecule found in nature that has a role in the control of the translation of pathogenesis-related proteins [71]. Apart from the defense responses, it is important in the regulation of plant development, ripening, and growth, and also in abiotic stress responses [72, 73]. The PAL (phenylalanine ammonia-lyase) and IC (isochorismate) pathways are both used to synthesize SA. In tomato [74] and *Nicotiana benthamiana* [75], the IC route is the most important. Low amounts of SA boost plant antioxidant capability, but large quantities of salicylic acid promote apoptosis or sensitivity to abiotic stressors [76]. Salicylic acid induces genes engaged in secondary metabolite formation, such as cytochrome P450, cinnamyl alcohol dehydrogenase and sinapyl alcohol dehydrogenase, antioxidants, heat shock proteins and chaperones [76]. Salicylic acid, alongside abscisic acid, is engaged in drought response control [77]. In *Phillyrea angustifolia*, drought stress caused a 5-fold rise in levels of endogenous salicylic acid [78]. Water scarcity raised the SA concentration in barley roots by about threefold [79]. Drought stress induces the salicylic acid-inducible genes *viz.*, pathogenesis-related protein-1 and pathogenesis-related protein-2 (PR2 and PR1), which are related to the Although, the specificpathogenesis Although, the specific [80]. Although, the specific molecular mechanisms underlying salicylic acid's functions in abiotic stress resilience are mostly unclear, and further research is required in this area [5].

2.9. Strigolactones (SL)

They are a small family of carotenoid-derived chemicals that were initially detected as seed germination signaling molecules in root parasite plants like Phelipanche, Orobanche, Striga and species more than 45 years ago [81, 82]. Species of a single plant can create a variety of strigolactones, whereas intraspecific variants produce mixtures of diverse quantities and types of strigolactones [82, 83]. Although they are largely generated and secreted in modest amounts in roots, they can also be synthesized in other plant sections [84]. Research comparing mutant and wild-type Arabidopsis plants found that they play a function in root system architecture development [85]. The use of GR24, biologically and synthetically active strigolactones [86, 87] suppressed the lateral formation of root in strigolactones-synthesis mutants and wild-type seedlings (max4 and max3) but not in the strigolactone-response mutant (max2), indicating that strigolactone hurts the lateral formation of root [86, 88]. From the beginning of their evolution, SLs appear to have been involved in plant responses to external stimuli. They engage in both root and shoot structures in higher plants in response to nutritional circumstances [89]. Strigolactones also serve as signaling molecules in interactions of the plant-microbe. In the legume–rhizobium interaction mechanism, they increase nodulation [90, 91]. In general, it can be inferred that SLs are an essential class of signaling molecules that play a crucial role in plants' developmental responses to varying environmental situations. They can be used in agriculture to meet many needs including as inducers of parasitic plants' suicidal seed germination [92]. The different roles played by key phytohormones to coordinate signal transduction pathways are summarised Table (**1**), and a summary of mitigation of abiotic stress in different crops due to the application of phytohormones is presented Table (**2**).

Table 1. Summary of key phytohormones: precursors, pathways for synthesis, site of synthesis and key functions.

Phytohormones	Biosynthetic Pathway	Precursor	Site of Synthesis	Functions
ABA	Carotenoid biosynthesis pathway	IPP-derived tetra-terpene (phytoene)	All main organs and living tissues of plants	Under water stress, the closing of stomata, modulation of sodium & potassium uptake in the guard cells, dormancy of seeds.

(Table 1) cont.....

Phytohormones	Biosynthetic Pathway	Precursor	Site of Synthesis	Functions
IAA	Tryptophan biosynthetic pathway	Tryptophan	Young leaves, shoot apical meristems (major), root apical meristems	Stimulation of cell expansion, formation of a bud, initiation of the root, along with cytokinins, growth of stems, roots, and fruits is controlled, conversion of stems into flowers. Role in gravitropism & phototropism.
CKs	Zeatin biosynthesis	Adenine	Root tissues (major)	Regulates cell division in root & shoot, stimulation and germination of seed & delaying of senescence, promotion of lateral growth of buds and modification of apical dominance.
BRs	Brassinosteroid biosynthetic pathway	Farnesyl diphosphate	All plant tissues	Control of cell elongation & division, gravitropism, differentiation of xylem, the resistance of stress, inhibition of root growth and abscission of leaves.
ET	Cysteine and methionine metabolism	Methionine	Most parts of the plant, rapidly grow and divide cells in dark conditions	Inhibition of stem elongation, promotion of lateral leaf expansion, horizontal growth of the seedling, enhancement of senescence rate, promotion of root and root hairs, and ripening fruits.
GAs	Isoprenoid pathway	Geranylgeranyl diphosphate	Meristematic tissues of emerging seeds, apical buds, new leaves and roots.	Seed germination, sprouting of buds, promotion of cell elongation, and alteration between vegetative and reproductive growth, are required for the functioning of pollen during fertilization and fruit development.
JA	Linolenic acid pathway/ octadecanoid pathway (lipid metabolism)	Linolenic acid	Synthesized in various parts and moved through the phloem to other parts of the plant.	Plant response against attack from herbivores and necrotrophic pathogens, germination of seed, storage of protein in seeds and root growth.
SA	Phenylalanine metabolism	Chorismate		Involved in defense against biotrophic pathogens as well as in abiotic stresses.

(Table 1) cont.....

Phytohormones	Biosynthetic Pathway	Precursor	Site of Synthesis	Functions
SLs	Carotenoid biosynthesis pathway	Carotenoids	Under low phosphate conditions, produced in root and under high auxin flow, produced in shoots	Senescence of leaves, response to phosphate starvation, tolerance to salt, signaling of light, inhibition of branching of the shoot.

Table 2. Phytohormone-mediated mitigation of abiotic stress tolerance.

Abiotic stress	Plant	Phytohormones applied	Effect of Phytohormone on plant	References
Salinity	*Zea mays* L.	Auxin	In saline conditions, increase in different types of antioxidant activities. Reduction in the concentration of sodium ion and improvement in K, Ca $^{2+}$ and P levels. Promotion of growth was linked with photosynthetic pigment content. Reduction of permeability of the membrane.	[93]
Salt	*Zea mays* L.	Auxin	Salt stress-induced reduction of photosynthesis and growth, but the application of IAA alleviated the reduction of growth & photosynthesis and concurrently, reduction in the accumulation of Na$^+$ in shoot and roots.	[94]
Cadmium	*Pisum sativum*	Salicyclic acid	SA alleviated the negative effect of Cd toxicity on growth, photosynthesis, carboxylation reactions, thermoluminescence characteristics, and chlorophyll content along with a decrease in oxidative injuries.	[95]
Copper	*Helianthus annus* L.	Auxin, Gibberellin	Application of hormones led to substantial preservation of chlorophylls and carotenoids, stability of the light-harvesting complex of PS2 reaction centers and net photosynthetic rate.	[96]
Zn	*Raphanus sativus* L.	24-Epibrassinolide (EBL)	EBL reduced oxidative stress by increasing the activity of guaiacol peroxidase, superoxide dismutase, and glutathione peroxidase.	[97]
Chilling stress	*Cucumis sativus* L.	28-HBL (28-Homobrassinolide)	Application of HBL increased the quantum yield of PS II, and improved photosynthesis, growth and water relations.	[98]

(Table 2) cont.....

Abiotic stress	Plant	Phytohormones applied	Effect of Phytohormone on plant	References
UV radiation	*Glycine max* L.	Epibrassinosteroid	Epibrassinosteroid could minimize the decrease of chlorophyll a, chlorophyll b & carotenoid content. Treatment of plants with epibrassinosteroid and UV-B/UV-C showed an increased level of UV absorbing compounds and pigments (anthocyanin and flavonoids).	[99]
Drought stress	*Glycine max* L.	Brassinolide	Application of BR augmented the maximum quantum yield of PS II, ribulose-1,5-bisphosphate carboxylase activity, and the leaf water potential of drought-stressed plants. Treatment also increased the concentration of soluble sugars and proline, and the activities of peroxidase and superoxide dismutase of soybean leaves when drought-stressed.	[100]
Drought stress	*Brassica juncea* L.	28-HBL	HBL restored growth and photosynthesis. Increment of activity of SOD, POD, CAT & proline.	[101]

3. MOLECULAR STUDIES AND METABOLIC ENGINEERING OF PHYTOHORMONES

Alterations in phytohormone and, consequently signal transduction cascade are the main responses to stress in plants [102]. Since, phytohormones are the chief regulators of growth and development as well as responses to environmental stresses, the signaling processes and hormone metabolism becomes a superb target for alteration to get augmented tolerance to abiotic stresses. Studies conducted so far over the past several decades identified several genes that are either associated with or affect biosynthesis, passage, metabolism and perception of phytohormones. Among many plant hormones, ABA is chosen for engineering for conferring tolerance to abiotic stresses in crop plants since it performs a wide range of roles under different stress conditions, mainly drought. Consequently, several enzymes in the biosynthetic pathway of ABA have been studied for conferring abiotic stress tolerance [103]. ABA-responsive stress gene was overexpressed in Arabidopsis, and as a result, osmotic stress tolerance was increased. When genes responsible for ABA synthesis or catabolic pathways were overexpressed, enhanced drought tolerance was achieved, but that led to defective growth owing to pleiotropic effects even after inducible promoters were used [104]. To avoid these growth defects, Cysteine Rich Receptor-like Kinase (CRK45), a kinase inducible in stress involved in ABA signaling, was overexpressed, and the developed transgenic plant had improved drought tolerance indicating CRK45 was able to fine-tune ABA levels [105]. Similarly, using stress-inducible promoters, IPT was expressed to evade pleiotropic effects that led to an increment in CK level, scavenging of antioxidants, and better root growth leading to enhancement in grain yield under drought conditions in *Agrostis stolonifera* [106, 107]. Similarly, the Arabidopsis YUCCA6 gene of the tryptophan-dependent IAA biosynthesis pathway was overexpressed, resulting in the development of transgenic poplars under the control of the SWPA2 promoter which is inducible in stress [108]. The transgenic plants showed morphological features like faster growth of shoots, stunted development of roots I and higher induction of root hairs. Similarly, *OsGA2ox1*, codes for GA2-oxidase, were when overexpressed in rice with the actin promoter, GA levels were modified that brought dwarfism and without setting of grain [5]. Various attempts for engineering phytohormones for higher tolerance of plants to abiotic stress are listed in Table (**3**).

Table 3. Phytohormone engineering efforts conferring abiotic stress tolerance in plants.

Phytohormone	Gene Involved	The Function of the Gene Involved	Transgenic Plant and Differential Expression Level	Tolerance Level and Phenotype of the Transgenic Plant(s)	Reference
ABA	MoCo sulfurase	Act as a regulator of the final step of ABA synthesis	Overexpressed in soybean	Increase in biomass yield and drought tolerance.	[109]
-	LOS5	Act as an important regulator of ABA biosynthesis	Overexpressed in maize	Increase in ABA accumulation & drought tolerance.	[110]
-	AtLOS5	Act as an important regulator of ABA biosynthesis	Overexpressed in maize	Increase in salinity tolerance.	[111]
-	NCED (9-cis-epoxycarotenoid dioxygenase)	Play a significant function in feedback control in the rate-limiting step of ABA biosynthesis	Overexpressed in petunia	Increment of endogenous ABA reduced conductance of stomata & enhanced tolerance to drought.	[112]
-	MsZEP (zeaxanthin epoxidase)	An important role in ABA biosynthesis	Overexpressed in tobacco	Tolerance to salt and drought is enhanced.	[113]

(Table 3) cont.....

Phytohormone	Gene Involved	The Function of the Gene Involved	Transgenic Plant and Differential Expression Level	Tolerance Level and Phenotype of the Transgenic Plant(s)	Reference
-	SnRK2.4 (sucrose non-fermentin--related kinase 2 family)	In the ABA signaling network, act as an important serine/threonine-protein kinase	Overexpressed in tobacco	Enhanced drought, salt and freezing tolerance are linked with reduced loss of water, better osmotic potential and photosynthesis.	[114]
-	OsPIN3t (auxin efflux carrier gene)	Act as a carrier of auxin efflux, significant in the transport of polar auxin	Overexpressed in rice	Enhanced drought tolerance.	[115]
Auxin	YUCCA6 (Tryptophan aminotransferase of *Arabidopsis* (TAA)/YUCCA)	Important gene in auxin/IPA biosynthesis	Overexpressed in poplar	Drought and oxidative stress, tolerance is achieved.	[116]
-	OsIAA6	A member of the IAA/rice auxin gene family	Overexpressed in rice	Enhanced drought tolerance	[117]
-	IPT (Isopentyl transferase gene)	Biosynthesis of cytokinin	Overexpressed in tomato	Salt tolerance is enhanced.	[118]
Cytokinins	CKX	Cytokinin dehydrogenase	Overexpressed in arabidopsis	Tolerance is enhanced for drought.	[119]
-	AtCKX1	Cytokinin dehydrogenase	Overexpressed in barley	Better drought tolerance.	[120]
-	ERF-1 (Ethylene response factor-1) (JERF1)	For ethylene and Jasmonates, it acts as a response factor	Overexpressed in rice	Drought tolerance is augmented.	[121]
Ethylene	ETOL1		Overexpressed in rice	Submergence & drought tolerance is increased.	[122]
-	ACC-Synthase (1-aminocyclopropane-1-carboxylic acid synthase)	Catalyzes rate-limiting step in ethylene biosynthesis	Gene silenced in maize	Improved drought tolerance.	[123]
-	ZmARGOS (Auxin-Regulated *Gene* Involved in Organ Size)	Regulates ethylene signal transduction negatively.	Overexpressed in Arabidopsis and maize	Improvement in drought tolerance.	[124]

(Table 3) cont.....

Phytohormone	Gene Involved	The Function of the Gene Involved	Transgenic Plant and Differential Expression Level	Tolerance Level and Phenotype of the Transgenic Plant(s)	Reference
-	*OsGSK1* (Os GLYCOGEN SYNTHASE KINASE 1)	Act as a negative regulator of BR	Knockout of *OsGSK1* in rice	Knockout mutants had improved tolerance to drought, salt, heat & cold.	[125]
Brassinosteroids	*AtHSD1* (Hydroxysteroid Dehydrogenase)	Role in BR biosynthesis	Overexpressed in Arabidopsis	Salinity tolerance, growth rate & seed yield is increased.	[126]
-	*BdBR1*	BR-receptor gene	Downregulation in purple false brome.	Drought tolerance is improved.	[127]

CONCLUSION AND PERSPECTIVES

The research attempts conducted so far have indicated that phytohormone engineering can be one of the feasible approaches for producing stress-tolerant cultivars to meet the growing demands of the burgeoning population under changing environmental conditions. Although with advancements in genomic research and genetic engineering technology, abiotic stress tolerance mechanisms have been deciphered, and consequently, transgenic plants with enhanced abiotic stress tolerance have been developed. However, many complexities of stress signal transduction pathways remain unexplored. There are still many gaps that need detailed study *viz.* understanding the role of different differentially regulated genes and enzymes involved in the biosynthetic pathway of various phytohormones as well as their developmental and physiological responses under stress conditions. Since there is a variable response of different phytohormones in response to abiotic stress, gene pyramiding under the regulation of suitable promoters can confer higher abiotic stress tolerance. The knowledge generated can be useful in unraveling complex transcriptional networks that can lead to identifying and characterizing novel candidate genes for developing crop plants with enhanced abiotic stress tolerance.

REFERENCES

[1]   Wani SH, Kumar Sah S. Biotechnology and abiotic stress tolerance in rice. Rice Research: Open Access 2014; 2(2).
[http://dx.doi.org/10.4172/jrr.1000e105]

[2]   Vardhini BV, Anjum NA. Brassinosteroids make plant life easier under abiotic stresses mainly by modulating major components of antioxidant defense system. Front Environ Sci 2015; 2: 1-16.

[http://dx.doi.org/10.3389/fenvs.2014.00067]

[3]   Tilman D, Balzer C, Hill J, Befort BL. Global food demand and the sustainable intensification of agriculture. Proc Natl Acad Sci USA 2011; 108(50): 20260-4.
[http://dx.doi.org/10.1073/pnas.1116437108] [PMID: 22106295]

[4]   Qin F, Shinozaki K, Yamaguchi-Shinozaki K. Achievements and challenges in understanding plant abiotic stress responses and tolerance. Plant Cell Physiol 2011; 52(9): 1569-82.
[http://dx.doi.org/10.1093/pcp/pcr106] [PMID: 21828105]

[5]   Wani SH, Kumar V, Shriram V, Sah SK. Phytohormones and their metabolic engineering for abiotic stress tolerance in crop plants. Crop J 2016; 4(3): 162-76.
[http://dx.doi.org/10.1016/j.cj.2016.01.010]

[6]   Voß U, Bishopp A, Farcot E, Bennett MJ. Modelling hormonal response and development. Trends Plant Sci 2014; 19(5): 311-9.
[http://dx.doi.org/10.1016/j.tplants.2014.02.004] [PMID: 24630843]

[7]   Kazan K. Diverse roles of jasmonates and ethylene in abiotic stress tolerance. Trends Plant Sci 2015; 20(4): 219-29.
[http://dx.doi.org/10.1016/j.tplants.2015.02.001] [PMID: 25731753]

[8]   Wolters H, Jürgens G. Survival of the flexible: hormonal growth control and adaptation in plant development. Nat Rev Genet 2009; 10(5): 305-17.
[http://dx.doi.org/10.1038/nrg2558] [PMID: 19360022]

[9]   Fahad S, Hussain S, Bano A, *et al.* Potential role of phytohormones and plant growth-promoting rhizobacteria in abiotic stresses: consequences for changing environment. Environ Sci Pollut Res Int 2015; 22(7): 4907-21.
[http://dx.doi.org/10.1007/s11356-014-3754-2] [PMID: 25369916]

[10]   Peleg Z, Blumwald E. Hormone balance and abiotic stress tolerance in crop plants. Curr Opin Plant Biol 2011; 14(3): 290-5.
[http://dx.doi.org/10.1016/j.pbi.2011.02.001] [PMID: 21377404]

[11]   Sreenivasulu N, Radchuk V, Alawady A, *et al.* De-regulation of abscisic acid contents causes abnormal endosperm development in the barley mutant seg8. Plant J 2010; 64(4): 589-603.
[http://dx.doi.org/10.1111/j.1365-313X.2010.04350.x] [PMID: 20822501]

[12]   O'Brien JA, Benková E. Cytokinin cross-talking during biotic and abiotic stress responses. Front Plant Sci 2013; 4: 451.
[http://dx.doi.org/10.3389/fpls.2013.00451] [PMID: 24312105]

[13]   Nemhauser JL, Hong F, Chory J. Different plant hormones regulate similar processes through largely nonoverlapping transcriptional responses. Cell 2006; 126(3): 467-75.
[http://dx.doi.org/10.1016/j.cell.2006.05.050] [PMID: 16901781]

[14]   Keskin BC, Sarikaya AT, Yuksel B, Memon AR. Abscisic acid regulated gene expression in bread wheat. Aust J Crop Sci 2010; 4: 617-25.

[15]   Wilkinson S, Kudoyarova GR, Veselov DS, Arkhipova TN, Davies WJ. Plant hormone interactions: innovative targets for crop breeding and management. J Exp Bot 2012; 63(9): 3499-509.
[http://dx.doi.org/10.1093/jxb/ers148] [PMID: 22641615]

[16]   Giuliani S, Sanguineti MC, Tuberosa R, Bellotti M, Salvi S, Landi P. Root-ABA1, a major constitutive QTL, affects maize root architecture and leaf ABA concentration at different water regimes. J Exp Bot 2005; 56(422): 3061-70.
[http://dx.doi.org/10.1093/jxb/eri303] [PMID: 16246858]

[17]   Zhang S, Hu J, Zhang Y, Xie XJ, Knapp A. Seed priming with brassinolide improves lucerne (Medicago sativa L.) seed germination and seedling growth in relation to physiological changes under salinity stress. Aust J Agric Res 2007; 58(8): 811-5.
[http://dx.doi.org/10.1071/AR06253]

[18]   Verslues PE, Agarwal M, Katiyar-Agarwal S, Zhu J, Zhu JK. Methods and concepts in quantifying resistance to drought, salt and freezing, abiotic stresses that affect plant water status. Plant J 2006; 45(4): 523-39.
[http://dx.doi.org/10.1111/j.1365-313X.2005.02593.x] [PMID: 16441347]

[19]   Sreenivasulu N, Harshavardhan VT, Govind G, Seiler C, Kohli A. Contrapuntal role of ABA: Does it mediate stress tolerance or plant growth retardation under long-term drought stress? Gene 2012; 506(2): 265-73.
[http://dx.doi.org/10.1016/j.gene.2012.06.076] [PMID: 22771691]

[20]   Chaves MM, Maroco JP, Pereira JS. Understanding plant responses to drought — from genes to the whole plant. Funct Plant Biol 2003; 30(3): 239-64.
[http://dx.doi.org/10.1071/FP02076] [PMID: 32689007]

[21]   Zhang J, Jia W, Yang J, Ismail AM. Role of ABA in integrating plant responses to drought and salt stresses. Field Crops Res 2006; 97(1): 111-9.
[http://dx.doi.org/10.1016/j.fcr.2005.08.018]

[22]   Ke Q, Wang Z, Ji CY, *et al.* Transgenic poplar expressing *Arabidopsis* YUCCA6 exhibits auxin-overproduction phenotypes and increased tolerance to abiotic stress. Plant Physiol Biochem 2015; 94: 19-27.
[http://dx.doi.org/10.1016/j.plaphy.2015.05.003] [PMID: 25980973]

[23]   Mano Y, Nemoto K. The pathway of auxin biosynthesis in plants. J Exp Bot 2012; 63(8): 2853-72.
[http://dx.doi.org/10.1093/jxb/ers091] [PMID: 22447967]

[24]   Kazan K. Auxin and the integration of environmental signals into plant root development. Ann Bot (Lond) 2013; 112(9): 1655-65.
[http://dx.doi.org/10.1093/aob/mct229] [PMID: 24136877]

[25]   De Smet I, Voß U, Lau S, *et al.* Unraveling the evolution of auxin signaling. Plant Physiol 2011; 155(1): 209-21.
[http://dx.doi.org/10.1104/pp.110.168161] [PMID: 21081694]

[26]   Iqbal N, Umar S, Khan NA, Khan MIR. A new perspective of phytohormones in salinity tolerance: Regulation of proline metabolism. Environ Exp Bot 2014; 100: 34-42.
[http://dx.doi.org/10.1016/j.envexpbot.2013.12.006]

[27]   Sheng XF, Xia JJ. Improvement of rape (*Brassica napus*) plant growth and cadmium uptake by cadmium-resistant bacteria. Chemosphere 2006; 64(6): 1036-42.
[http://dx.doi.org/10.1016/j.chemosphere.2006.01.051] [PMID: 16516946]

[28]   Egamberdieva D. Alleviation of salt stress by plant growth regulators and IAA producing bacteria in wheat. Acta Physiol Plant 2009; 31(4): 861-4.
[http://dx.doi.org/10.1007/s11738-009-0297-0]

[29]   Javid MG, Sorooshzadeh A, Moradi F, Sanavy SAMM, Allahdadi I. The role of phytohormones in alleviating salt stress in crop plants. Aust J Crop Sci 2011; 5: 726-34.

[30]   Fahad S, Hussain S, Matloob A, *et al.* Phytohormones and plant responses to salinity stress: a review. Plant Growth Regul 2015; 75(2): 391-404.
[http://dx.doi.org/10.1007/s10725-014-0013-y]

[31]   Kang NY, Cho C, Kim NY, Kim J. Cytokinin receptor-dependent and receptor-independent pathways in the dehydration response of *Arabidopsis thaliana*. J Plant Physiol 2012; 169(14): 1382-91.
[http://dx.doi.org/10.1016/j.jplph.2012.05.007] [PMID: 22704545]

[32]   Nishiyama R, Watanabe Y, Fujita Y, *et al.* Analysis of cytokinin mutants and regulation of cytokinin metabolic genes reveals important regulatory roles of cytokinins in drought, salt and abscisic acid responses, and abscisic acid biosynthesis. Plant Cell 2011; 23(6): 2169-83.
[http://dx.doi.org/10.1105/tpc.111.087395] [PMID: 21719693]

[33] Zalabák D, Pospíšilová H, Šmehilová M, Mrízová K, Frébort I, Galuszka P. Genetic engineering of cytokinin metabolism: Prospective way to improve agricultural traits of crop plants. Biotechnol Adv 2013; 31(1): 97-117.
[http://dx.doi.org/10.1016/j.biotechadv.2011.12.003] [PMID: 22198203]

[34] Pospíšilová J. Interaction of cytokinins and abscisic acid during regulation of stomatal opening in bean leaves. Photosynthetica 2003; 41(1): 49-56.
[http://dx.doi.org/10.1023/A:1025852210937]

[35] Pospíšilová J. Participation of phytohormones in the stomatal regulation of gas exchange during water stress. Biol Plant 2003; 46(4): 491-506.
[http://dx.doi.org/10.1023/A:1024894923865]

[36] Gamalero E, Glick BR. Ethylene and abiotic stress tolerance in plants.Environmental Adaptations and Stress Tolerance of Plants in the Era of Climate Change. New York: Springer 2012; pp. 395-412.
[http://dx.doi.org/10.1007/978-1-4614-0815-4_18]

[37] Groen SC, Whiteman NK. The evolution of ethylene signaling in plant chemical ecology. J Chem Ecol 2014; 40(7): 700-16.
[http://dx.doi.org/10.1007/s10886-014-0474-5] [PMID: 24997626]

[38] Shi Y, Tian S, Hou L, et al. Ethylene signaling negatively regulates freezing tolerance by repressing expression of CBF and type-A ARR genes in *Arabidopsis*. Plant Cell 2012; 24(6): 2578-95.
[http://dx.doi.org/10.1105/tpc.112.098640] [PMID: 22706288]

[39] Larkindale J, Hall JD, Knight MR, Vierling E. Heat stress phenotypes of Arabidopsis mutants implicate multiple signaling pathways in the acquisition of thermotolerance. Plant Physiol 2005; 138(2): 882-97.
[http://dx.doi.org/10.1104/pp.105.062257] [PMID: 15923322]

[40] Klay I, Pirrello J, Riahi l, et al. Ethylene response factor Sl-ERF.B.3 is responsive to abiotic stresses and mediates salt and cold stress response regulation in tomato. Sci World J 2014; 2014: 167681.
[http://dx.doi.org/10.1155/2014/167681]

[41] Matilla-Vazquez MA, Matilla AJ. Matilla, Ethylene: Role in plants under environmental stress, in: P. Ahmad, M.R. Wani (Eds.), Physiological Mechanisms and Adaptation Strategies in Plants under Changing Environment, Springer Science + Business Media: New York 2014; 2, pp. 189–222.

[42] Yin CC, Ma B, Collinge DP, et al. Ethylene responses in rice roots and coleoptiles are differentially regulated by a carotenoid isomerase-mediated abscisic acid pathway. Plant Cell 2015; 27(4): 1061-81.
[http://dx.doi.org/10.1105/tpc.15.00080] [PMID: 25841037]

[43] Sponsel VM, Hedden P. Gibberellin, biosynthesis and inactivation.Plant Hormones Biosynthesis, Signal Transduction, Action!. Dordrecht: Springer 2004; pp. 63-94.

[44] Yamaguchi S. Gibberellin metabolism and its regulation. Annu Rev Plant Biol 2008; 59(1): 225-51.
[http://dx.doi.org/10.1146/annurev.arplant.59.032607.092804] [PMID: 18173378]

[45] Colebrook EH, Thomas SG, Phillips AL, Hedden P. The role of gibberellin signalling in plant responses to abiotic stress. J Exp Biol 2014; 217(1): 67-75.
[http://dx.doi.org/10.1242/jeb.089938] [PMID: 24353205]

[46] Skirycz A, Claeys H, De Bodt S, et al. Pause-and-stop: the effects of osmotic stress on cell proliferation during early leaf development in Arabidopsis and a role for ethylene signaling in cell cycle arrest. Plant Cell 2011; 23(5): 1876-88.
[http://dx.doi.org/10.1105/tpc.111.084160] [PMID: 21558544]

[47] Claeys H, Skirycz A, Maleux K, Inzé D. DELLA signaling mediates stress-induced cell differentiation in *Arabidopsis* leaves through modulation of anaphase-promoting complex/cyclosome activity. Plant Physiol 2012; 159(2): 739-47.
[http://dx.doi.org/10.1104/pp.112.195032] [PMID: 22535421]

[48]   Munteanu V, Gordeev V, Martea R, Duca M. Effect of gibberellin cross talk with other phytohormones on cellular growth and mitosis to endoreduplication transition. Int J Adv Res Biol Sci 2014; 1(6): 136-53.

[49]   Vardhini BV, Anuradha S, Rao SS. Brassinosteroids-New class of plant hormone with potential to improve crop productivity. Indian J Plant Physiol 2006; 11(1): 1.

[50]   Bajguz A, Hayat S. Effects of brassinosteroids on the plant responses to environmental stresses. Plant Physiol Biochem 2009; 47(1): 1-8.
[http://dx.doi.org/10.1016/j.plaphy.2008.10.002] [PMID: 19010688]

[51]   Ahammed GJ, Choudhary SP, Chen S, *et al.* Role of brassinosteroids in alleviation of phenanthrene–cadmium co-contamination-induced photosynthetic inhibition and oxidative stress in tomato. J Exp Bot 2013; 64(1): 199-213.
[http://dx.doi.org/10.1093/jxb/ers323] [PMID: 23201830]

[52]   Bajguz A. An enhancing effect of exogenous brassinolide on the growth and antioxidant activity in Chlorella vulgaris cultures under heavy metals stress. Environ Exp Bot 2010; 68(2): 175-9.
[http://dx.doi.org/10.1016/j.envexpbot.2009.11.003]

[53]   Liang JQ, Liang Y. Effects of plant growth substances on water-logging resistance of oilseed rape seedling J Southwest China Norm Univ (Nat Sci Ed) . 2009; 34: pp. 58-62. (in Chinese with English abstract).

[54]   Mahesh K, Balaraju P, Ramakrishna B, Ram Rao SS. Effect of brassinosteroids on germination and seedling growth of radish (Raphanus sativus L.) under PEG-6000 induced water stress. Am J Plant Sci 2013; 4(12): 2305-13.
[http://dx.doi.org/10.4236/ajps.2013.412285]

[55]   Kurepin LV, Joo SH, Kim SK, Pharis RP, Back TG. Interaction of brassinosteroids with light quality and plant hormones in regulating shoot growth of young sunflower and Arabidopsis seedlings. J Plant Growth Regul 2012; 31(2): 156-64.
[http://dx.doi.org/10.1007/s00344-011-9227-7]

[56]   Abbas S, Latif HH, Elsherbiny EA. Effect of 24-epibrassinolide on the physiological and genetic changes on two varieties of pepper under salt stress conditions. Pak J Bot 2013; 45(4): 1273-84.

[57]   Wang XH, Shu C, Li HY, Hu XQ, Wang YX. Effects of 0.01% brassinolide solution application on yield of rice and its resistance to autumn low-temperature damage. Jiangxi Nongye Daxue Xuebao 2014; 26: 36-8. [in Chinese with English abstract].

[58]   Janeczko A, Oklešťková J, Pociecha E, Kościelniak J, Mirek M. Physiological effects and transport of 24-epibrassinolide in heat-stressed barley. Acta Physiol Plant 2011; 33(4): 1249-59.
[http://dx.doi.org/10.1007/s11738-010-0655-y]

[59]   Seo HS, Song JT, Cheong JJ, *et al.* Jasmonic acid carboxyl methyltransferase: A key enzyme for jasmonate-regulated plant responses. Proc Natl Acad Sci USA 2001; 98(8): 4788-93.
[http://dx.doi.org/10.1073/pnas.081557298] [PMID: 11287667]

[60]   Fahad S, Nie L, Chen Y, *et al.* Crop plant hormones and environmental stress. Sustainable agriculture reviews 2015; 371-400.

[61]   Pauwels L, Inzé D, Goossens A. Jasmonate-inducible gene: what does it mean? Trends Plant Sci 2009; 14(2): 87-91.
[http://dx.doi.org/10.1016/j.tplants.2008.11.005] [PMID: 19162528]

[62]   Seo JS, Joo J, Kim MJ, *et al.* OsbHLH148, a basic helix-loop-helix protein, interacts with OsJAZ proteins in a jasmonate signaling pathway leading to drought tolerance in rice. Plant J 2011; 65(6): 907-21.

[63]   Demkura PV, Abdala G, Baldwin IT. Ballaren CL. Jasmonate-dependent and-independent pathways mediate specific effects of solar ultraviolet B radiation on leaf phenolics and antiherbivore defense.

Plant Physiol 2010; 152(2): 1084-95.
[http://dx.doi.org/10.1104/pp.109.148999] [PMID: 20007446]

[64] Seo JS, Joo J, Kim MJ, *et al.* OsbHLH148, a basic helix-loop-helix protein, interacts with OsJAZ proteins in a jasmonate signaling pathway leading to drought tolerance in rice. Plant J 2011; 65(6): 907-21.
[http://dx.doi.org/10.1111/j.1365-313X.2010.04477.x] [PMID: 21332845]

[65] Du H, Liu H, Xiong L. Endogenous auxin and jasmonic acid levels are differentially modulated by abiotic stresses in rice. Front Plant Sci 2013; 4: 397.
[http://dx.doi.org/10.3389/fpls.2013.00397] [PMID: 24130566]

[66] Dar TA, Uddin M, Khan MMA, Hakeem KR, Jaleel H. Jasmonates counter plant stress: A Review. Environ Exp Bot 2015; 115: 49-57.
[http://dx.doi.org/10.1016/j.envexpbot.2015.02.010]

[67] Yoon JY, Hamayun M, Lee SK, Lee IJ. Methyl jasmonate alleviated salinity stress in soybean. J Crop Sci Biotechnol 2009; 12(2): 63-8.
[http://dx.doi.org/10.1007/s12892-009-0060-5]

[68] Wang Y, Mopper S, Hasenstein KH. Effects of salinity on endogenous ABA, IAA, JA, AND SA in *Iris hexagona*. J Chem Ecol 2001; 27(2): 327-42.
[http://dx.doi.org/10.1023/A:1005632506230] [PMID: 14768818]

[69] Yan Z, Chen J, Li X. Methyl jasmonate as modulator of Cd toxicity in Capsicum frutescens var. fasciculatum seedlings. Ecotoxicol Environ Saf 2013; 98: 203-9.
[http://dx.doi.org/10.1016/j.ecoenv.2013.08.019] [PMID: 24064260]

[70] Maksymiec W, Wójcik M, Krupa Z. Variation in oxidative stress and photochemical activity in *Arabidopsis thaliana* leaves subjected to cadmium and excess copper in the presence or absence of jasmonate and ascorbate. Chemosphere 2007; 66(3): 421-7.
[http://dx.doi.org/10.1016/j.chemosphere.2006.06.025] [PMID: 16860844]

[71] Miura K, Tada Y. Regulation of water, salinity, and cold stress responses by salicylic acid. Front Plant Sci 2014; 5: 4.
[http://dx.doi.org/10.3389/fpls.2014.00004] [PMID: 24478784]

[72] Rivas-San Vicente M, Plasencia J. Salicylic acid beyond defence: its role in plant growth and development. J Exp Bot 2011; 62(10): 3321-38.
[http://dx.doi.org/10.1093/jxb/err031] [PMID: 21357767]

[73] Hara M, Furukawa J, Sato A, Mizoguchi T, Miura K. Abiotic stress and role of salicylic acid in plants 2012.
[http://dx.doi.org/10.1007/978-1-4614-0634-1_13]

[74] Catinot J, Buchala A, Abou-Mansour E, Métraux JP. Salicylic acid production in response to biotic and abiotic stress depends on isochorismate in *Nicotiana benthamiana*. FEBS Lett 2008; 582(4): 473-8.
[http://dx.doi.org/10.1016/j.febslet.2007.12.039] [PMID: 18201575]

[75] Uppalapati SR, Ishiga Y, Wangdi T, *et al.* The phytotoxin coronatine contributes to pathogen fitness and is required for suppression of salicylic acid accumulation in tomato inoculated with *Pseudomonas syringae* pv. tomato DC3000. Mol Plant Microbe Interact 2007; 20(8): 955-65.
[http://dx.doi.org/10.1094/MPMI-20-8-0955] [PMID: 17722699]

[76] Jumali SS, Said IM, Ismail I, Zainal Z. Genes induced by high concentration of salicylic acid in 'Mitragyna speciosa'. Aust J Crop Sci 2011; 5(3): 296-303.

[77] Miura K, Tada Y. Regulation of water, salinity, and cold stress responses by salicylic acid. Front Plant Sci 2014; 5: 4.
[http://dx.doi.org/10.3389/fpls.2014.00004] [PMID: 24478784]

[78] Munné-Bosch S, Peñuelas J. Photo- and antioxidative protection, and a role for salicylic acid during

drought and recovery in field-grown *Phillyrea angustifolia* plants. Planta 2003; 217(5): 758-66.
[http://dx.doi.org/10.1007/s00425-003-1037-0] [PMID: 12698367]

[79]   Bandurska H, Stroi ski A. The effect of salicylic acid on barley response to water deficit. Acta Physiol
        Plant 2005; 27(3): 379-86.
        [http://dx.doi.org/10.1007/s11738-005-0015-5]

[80]   Miura K, Okamoto H, Okuma E, *et al. SIZ1* deficiency causes reduced stomatal aperture and enhanced
        drought tolerance *via* controlling salicylic acid-induced accumulation of reactive oxygen species in
        Arabidopsis. Plant J 2013; 73(1): 91-104.
        [http://dx.doi.org/10.1111/tpj.12014] [PMID: 22963672]

[81]   Xie X, Yoneyama K, Yoneyama K. The strigolactone story. Annu Rev Phytopathol 2010; 48(1): 93-
        117.
        [http://dx.doi.org/10.1146/annurev-phyto-073009-114453] [PMID: 20687831]

[82]   Ruyter-Spira C, Al-Babili S, van der Krol S, Bouwmeester H. The biology of strigolactones. Trends
        Plant Sci 2013; 18(2): 72-83.
        [http://dx.doi.org/10.1016/j.tplants.2012.10.003] [PMID: 23182342]

[83]   Yoneyama K, Kisugi T, Xie X, Yoneyama K. Chemistry of strigolactones: why and how do plants
        produce so many strigolactones?Molecular Microbial Ecology of the Rhizosphere. Hoboken, NJ: John
        Wiley & Sons Ltd. 2013; Vol. 1 & 2: pp. 373-9.
        [http://dx.doi.org/10.1002/9781118297674.ch34]

[84]   Koltai H, Beveridge CA. Strigolactones and the coordinated development of shoot and root.Long-
        Distance Systemic Signaling and Communication in Plants. Berlin: Springer 2013; pp. 189-204.
        [http://dx.doi.org/10.1007/978-3-642-36470-9_9]

[85]   Ruyter-Spira C, Kohlen W, Charnikhova T, *et al.* Physiological effects of the synthetic strigolactone
        analog GR24 on root system architecture in *Arabidopsis*: another belowground role for strigolactones?
        Plant Physiol 2011; 155(2): 721-34.
        [http://dx.doi.org/10.1104/pp.110.166645] [PMID: 21119044]

[86]   Gomez-Roldan V, Fermas S, Brewer PB, *et al.* Strigolactone inhibition of shoot branching. Nature
        2008; 455(7210): 189-94.
        [http://dx.doi.org/10.1038/nature07271] [PMID: 18690209]

[87]   Umehara M, Hanada A, Yoshida S, *et al.* Inhibition of shoot branching by new terpenoid plant
        hormones. Nature 2008; 455(7210): 195-200.
        [http://dx.doi.org/10.1038/nature07272] [PMID: 18690207]

[88]   Kapulnik Y, Delaux PM, Resnick N, *et al.* Strigolactones affect lateral root formation and root-hair
        elongation in *Arabidopsis*. Planta 2011; 233(1): 209-16.
        [http://dx.doi.org/10.1007/s00425-010-1310-y] [PMID: 21080198]

[89]   Kapulnik Y, Koltai H. Strigolactone involvement in root development, response to abiotic stress, and
        interactions with the biotic soil environment. Plant Physiol 2014; 166(2): 560-9.
        [http://dx.doi.org/10.1104/pp.114.244939] [PMID: 25037210]

[90]   Soto MJ, Fernández-Aparicio M, Castellanos-Morales V, *et al.* First indications for the involvement of
        strigolactones on nodule formation in alfalfa (Medicago sativa). Soil Biol Biochem 2010; 42(2): 383-
        5.
        [http://dx.doi.org/10.1016/j.soilbio.2009.11.007]

[91]   Foo E, Davies NW. Strigolactones promote nodulation in pea. Planta 2011; 234(5): 1073-81.
        [http://dx.doi.org/10.1007/s00425-011-1516-7] [PMID: 21927948]

[92]   Vurro M, Yoneyama K. Strigolactones-intriguing biologically active compounds: perspectives for
        deciphering their biological role and for proposing practical application. Pest Manag Sci 2012; 68(5):
        664-8.
        [http://dx.doi.org/10.1002/ps.3257] [PMID: 22323399]

[93]   Egamberdieva D. Alleviation of salt stress by plant growth regulators and IAA producing bacteria in wheat. Acta Physiol Plant 2009; 31(4): 861-4.
[http://dx.doi.org/10.1007/s11738-009-0297-0]

[94]   Khalid S, Parvaiz M, Nawaz K, *et al.* Effect of Indole Acetic Acid (IAA) on morphological, biochemical and chemical attributes of Two varieties of maize (*Zea mays* L.) under salt stress. World Appl Sci J 2013; 26: 1150-9.

[95]   Popova LP, Maslenkova LT, Yordanova RY, *et al.* Exogenous treatment with salicylic acid attenuates cadmium toxicity in pea seedlings. Plant Physiol Biochem 2009; 47(3): 224-31.
[http://dx.doi.org/10.1016/j.plaphy.2008.11.007] [PMID: 19091585]

[96]   Ouzounidou G, Ilias I. Hormone-induced protection of sunflower photosynthetic apparatus against copper toxicity. Biol Plant 2005; 49(2): 223-8.
[http://dx.doi.org/10.1007/s10535-005-3228-y]

[97]   Ramakrishna B, Rao SSR. 24-Epibrassinolide alleviated zinc-induced oxidative stress in radish (*Raphanus sativus* L.) seedlings by enhancing antioxidative system. Plant Growth Regul 2012; 68(2): 249-59.
[http://dx.doi.org/10.1007/s10725-012-9713-3]

[98]   Fariduddin Q, Yusuf M, Chalkoo S, Hayat S, Ahmad A. 28-homobrassinolide improves growth and photosynthesis in *Cucumis sativus* L. through an enhanced antioxidant system in the presence of chilling stress. Photosynthetica 2011; 49(1): 55-64.
[http://dx.doi.org/10.1007/s11099-011-0022-2]

[99]   Enteshari S, Kalantari K, Ghorbani M. The effect of epibrassinosteroid and different bands of ultra violent radiation on the pigments content in glycine max. Pak J Biol Sci 2006; 9: 231-7.
[http://dx.doi.org/10.3923/pjbs.2006.231.237]

[100]   Zhang M, Zhai Z, Tian X, Duan L, Li Z. Brassinolide alleviated the adverse effect of water deficits on photosynthesis and the antioxidant of soybean (Glycine max L.). Plant Growth Regul 2008; 56(3): 257-64.
[http://dx.doi.org/10.1007/s10725-008-9305-4]

[101]   Fariduddin Q, Yusuf M, Hayat S, Ahmad A. Effect of 28-homobrassinolide on antioxidant capacity and photosynthesis in *Brassica juncea* plants exposed to different levels of copper. Environ Exp Bot 2009; 66(3): 418-24.
[http://dx.doi.org/10.1016/j.envexpbot.2009.05.001]

[102]   Vinay K, Saroj KS, Tushar K, Varsha S, Shabir HW. Engineering Phytohormones for Abiotic Stress Tolerance in Crop Plants.Plant hormones under challenging environmental factors. Springer-Dordrecht 2016; pp. 247-66.

[103]   Jewell MC, Campbell BC, Godwin ID. Transgenic plants for abiotic stress resistance, in: C. Kole, C. Michler, A.G. Abbott, T.C. Hall (Eds.), Transgenic Crop Plants, Utilization and Biosafety, Springer-Verlag: Berlin, 2010; 2, 67–131.
[http://dx.doi.org/10.1007/978-3-642-04812-8_2]

[104]   Hwang SG, Chen HC, Huang WY, Chu YC, Shii CT, Cheng WH. Ectopic expression of rice OsNCED3 in *Arabidopsis* increases ABA level and alters leaf morphology. Plant Sci 2010; 178(1): 12-22.
[http://dx.doi.org/10.1016/j.plantsci.2009.09.014] [PMID: 21421342]

[105]   Zhang X, Yang G, Shi R, *et al.* Arabidopsis cysteine-rich receptor-like kinase 45 functions in the responses to abscisic acid and abiotic stresses. Plant Physiol Biochem 2013; 67: 189-98.
[http://dx.doi.org/10.1016/j.plaphy.2013.03.013] [PMID: 23583936]

[106]   Xu Y, Burgess P, Zhang X, Huang B. Enhancing cytokinin synthesis by overexpressing *ipt* alleviated drought inhibition of root growth through activating ROS-scavenging systems in *Agrostis stolonifera*. J Exp Bot 2016; 67(6): 1979-92.

[http://dx.doi.org/10.1093/jxb/erw019] [PMID: 26889010]

[107] Merewitz EB, Du H, Yu W, Liu Y, Gianfagna T, Huang B. Elevated cytokinin content in ipt transgenic creeping bentgrass promotes drought tolerance through regulating metabolite accumulation. J Exp Bot 2012; 63(3): 1315-28.
[http://dx.doi.org/10.1093/jxb/err372] [PMID: 22131157]

[108] Ke Q, Wang Z, Ji CY, *et al.* Transgenic poplar expressing Arabidopsis YUCCA6 exhibits auxin-overproduction phenotypes and increased tolerance to abiotic stress. Plant Physiol Biochem 2015; 94: 19-27.
[http://dx.doi.org/10.1016/j.plaphy.2015.05.003] [PMID: 25980973]

[109] Li Y, Zhang J, Zhang J, *et al.* Expression of an *Arabidopsis* molybdenum cofactor sulphurase gene in soybean enhances drought tolerance and increases yield under field conditions. Plant Biotechnol J 2013; 11(6): 747-58.
[http://dx.doi.org/10.1111/pbi.12066] [PMID: 23581509]

[110] Lu Y, Li Y, Zhang J, *et al.* Overexpression of *Arabidopsis* molybdenum cofactor sulfurase gene confers drought tolerance in maize (*Zea mays* L.). PLoS One 2013; 8(1)e52126.
[http://dx.doi.org/10.1371/journal.pone.0052126] [PMID: 23326325]

[111] Zhang J, Yu H, Zhang Y, *et al.* Increased abscisic acid levels in transgenic maize overexpressing *AtLOS5* mediated root ion fluxes and leaf water status under salt stress. J Exp Bot 2016; 67(5): 1339-55.
[http://dx.doi.org/10.1093/jxb/erv528] [PMID: 26743432]

[112] Estrada-Melo AC, Chao , Reid MS, Jiang CZ. Overexpression of an ABA biosynthesis gene using a stress-inducible promoter enhances drought resistance in petunia. Hortic Res 2015; 2(1): 15013.
[http://dx.doi.org/10.1038/hortres.2015.13] [PMID: 26504568]

[113] Zhang ZQ, Wang YF, Chang LQ, *et al.* MsZEP, a novel zeaxanthin epoxidase gene from alfalfa (*Medicago sativa*), confers drought and salt tolerance in transgenic tobacco. Plant Cell Rep 2015; 14: 1-5.
[PMID: 26573680]

[114] Mao X, Zhang H, Tian S, Chang X, Jing R. TaSnRK2.4, an SNF1-type serine/threonine protein kinase of wheat (Triticum aestivum L.), confers enhanced multistress tolerance in Arabidopsis. J Exp Bot 2010; 61(3): 683-96.
[http://dx.doi.org/10.1093/jxb/erp331] [PMID: 20022921]

[115] Zhang Q, Li J, Zhang W, *et al.* The putative auxin efflux carrier *OsPIN3t* is involved in the drought stress response and drought tolerance. Plant J 2012; 72(5): 805-16.
[http://dx.doi.org/10.1111/j.1365-313X.2012.05121.x] [PMID: 22882529]

[116] Ke Q, Wang Z, Ji CY, *et al.* Transgenic poplar expressing *Arabidopsis* YUCCA6 exhibits auxin-overproduction phenotypes and increased tolerance to abiotic stress. Plant Physiol Biochem 2015; 94: 19-27.
[http://dx.doi.org/10.1016/j.plaphy.2015.05.003] [PMID: 25980973]

[117] Jung H, Lee DK, Choi YD, Kim JK. OsIAA6, a member of the rice Aux/IAA gene family, is involved in drought tolerance and tiller outgrowth. Plant Sci 2015; 236: 304-12.
[http://dx.doi.org/10.1016/j.plantsci.2015.04.018] [PMID: 26025543]

[118] Ghanem ME, Albacete A, Smigocki AC, *et al.* Root-synthesized cytokinins improve shoot growth and fruit yield in salinized tomato (*Solanum lycopersicum* L.) plants. J Exp Bot 2011; 62(1): 125-40.
[http://dx.doi.org/10.1093/jxb/erq266] [PMID: 20959628]

[119] Werner T, Nehnevajova E, Köllmer I, *et al.* Root-specific reduction of cytokinin causes enhanced root growth, drought tolerance, and leaf mineral enrichment in *Arabidopsis* and tobacco. Plant Cell 2011; 22(12): 3905-20.
[http://dx.doi.org/10.1105/tpc.109.072694] [PMID: 21148816]

[120]   Pospíšilová H, Jiskrová E, Vojta P, *et al.* Transgenic barley overexpressing a cytokinin dehydrogenase gene shows greater tolerance to drought stress. N Biotechnol 2016; 33(5): 692-705.
[http://dx.doi.org/10.1016/j.nbt.2015.12.005] [PMID: 26773738]

[121]   Zhang Z, Li F, Li D, Zhang H, Huang R. Expression of ethylene response factor JERF1 in rice improves tolerance to drought. Planta 2010; 232(3): 765-74.
[http://dx.doi.org/10.1007/s00425-010-1208-8] [PMID: 20574667]

[122]   Du H, Wu N, Cui F, You L, Li X, Xiong L. A homolog of ETHYLENE OVERPRODUCER, O s ETOL 1, differentially modulates drought and submergence tolerance in rice. Plant J 2014; 78(5): 834-49.
[http://dx.doi.org/10.1111/tpj.12508] [PMID: 24641694]

[123]   Habben JE, Bao X, Bate NJ, *et al.* Transgenic alteration of ethylene biosynthesis increases grain yield in maize under field drought-stress conditions. Plant Biotechnol J 2014; 12(6): 685-93.
[http://dx.doi.org/10.1111/pbi.12172] [PMID: 24618117]

[124]   Shi J, Habben JE, Archibald RL, *et al.* Overexpression of *ARGOS* genes modifies plant sensitivity to ethylene, leading to improved drought tolerance in both *Arabidopsis* and maize. Plant Physiol 2015; 169(1): 266-82.
[http://dx.doi.org/10.1104/pp.15.00780] [PMID: 26220950]

[125]   Koh S, Lee SC, Kim MK, *et al.* T-DNA tagged knockout mutation of rice OsGSK1, an orthologue of *Arabidopsis* BIN2, with enhanced tolerance to various abiotic stresses. Plant Mol Biol 2007; 65(4): 453-66.
[http://dx.doi.org/10.1007/s11103-007-9213-4] [PMID: 17690841]

[126]   Li F, Asami T, Wu X, Tsang EWT, Cutler AJ. A putative hydroxysteroid dehydrogenase involved in regulating plant growth and development. Plant Physiol 2007; 145(1): 87-97.
[http://dx.doi.org/10.1104/pp.107.100560] [PMID: 17616511]

[127]   Feng Y, Yin Y, Fei S. Down-regulation of BdBRI1, a putative brassinosteroid receptor gene produces a dwarf phenotype with enhanced drought tolerance in *Brachypodium distachyon.* Plant Sci 2015; 234: 163-73.
[http://dx.doi.org/10.1016/j.plantsci.2015.02.015] [PMID: 25804819]

CHAPTER 7

Living with Abiotic Stress from a Plant Nutrition Perspective in Arid and Semi-arid Regions

Nesreen H. Abou-Baker[1,*]

[1] *Soils and Water Use Department, National Research Centre, Dokki, Giza, Egypt*

Abstract: Mitigating the negative impacts of abiotic stress is an important approach, especially if climate change scenarios are realized. It is important to develop modern applications to deliver adequate and safe food for human consumption, particularly in arid and semi-arid regions that suffer from environmental and economic stressors. The progress made by scientific research in the field of plant tolerance to stress conditions during the last decade is considerable, but it needs to supply technical support for the application. The development strategy is based on combining more than one technique to achieve the integrated management of plants under different abiotic stresses, as will be described in this chapter.

Keywords: Abiotic stress, Management, Nano-technology, Plant, Soil, Water.

1. INTRODUCTION

The food gap is increasing with the severe increase in population density (the global population will be around 10 billion in 2050) in concomitant to the decrease in food production induced by environmental stress.

Environmental factors include natural forces such as the air, water, land, minerals, and all other external factors that affect a given living organism and influence its life for development and growth, as well as danger and damage at any time. The environment can also be defined as I) everything that is around us, II) the sum of all surroundings; circumstances, conditions, people, animals, plants, things, also abiotic and biotic factors that affect an organism in a particular geographical area.

[*] **Corresponding author Nesreen H. Abou-Baker :** Soils and Water Use Department, National Research Centre, Dokki, Giza, Egypt; Email: nesreenhaa@yahoo.com

Jen-Tsung Chen (Ed.)
All rights reserved-© 2023 Bentham Science Publishers

Before talking about the environment, three important factors should be determined; I) which kind of environment; natural, biophysical, historical or social environment, *etc.*, II) the studied living organisms such as humans, animals, or plants, and III) the specific geographical region; humidity or arid region, *etc.*

Environmental stress is defined as the pressure caused by natural events (Drought) or by peoples' activities (Pollution). Natural environmental stress (Ecological stress) is divided into two main types; biotic (The negative impact induced by living organisms such as parasites and pathogens; bacteria, viruses, nematodes, fungi, or insects) and abiotic (the damage caused by non-living factors) stress. The ability of plants to resist and survive under the stress is termed living with stress.

As well known, most of the newly reclaimed soils in Egypt are affected by more than one ecological stressor. Therefore, this work aims to give an overview of the modern techniques in living with abiotic stress as an important part of natural environmental stress and alleviate its impact on plant production under arid and semi-arid conditions.

2. BACKGROUND AND REVIEW OF LITERATURE

2.1. The Ecological Factors Related to Plant Production

The ecological factors are the summations of different components that affect plant survival, human health, and the economy. These factors can be classified into biotic factors that are related to living organisms (plants, animals, insects, and microorganisms) and abiotic factors which related to climate (temperature, light, precipitation, humidity, wind, and gases), topography (altitude, slope), soil, water, nutrients, *etc.*

The unfavorable ecological pressures are called ecological stressors. Only 3.5% of the global lands are not affected by ecological stress, according to the FAO report 2007 [1].

The ecological stressors are split into two primary kinds; biotic (pathogens and pests) and abiotic stress. Salinity (water or soil), water stress (drought or flooded), temperature (high or low), light (high or low) in addition to soil, water, or atmospheric pollution (heavy metals HMs, organic contaminants, high radiation, sound, magnetic and electrical pollution) caused by worse agriculture practices like unbalanced fertilization and excessive pesticide, the industry's wastes and fires which produced nitrogen oxides, hydrocarbons, hydrogen chloride, sulfur dioxide and dust as very dangerous contaminants Fig (**1**).

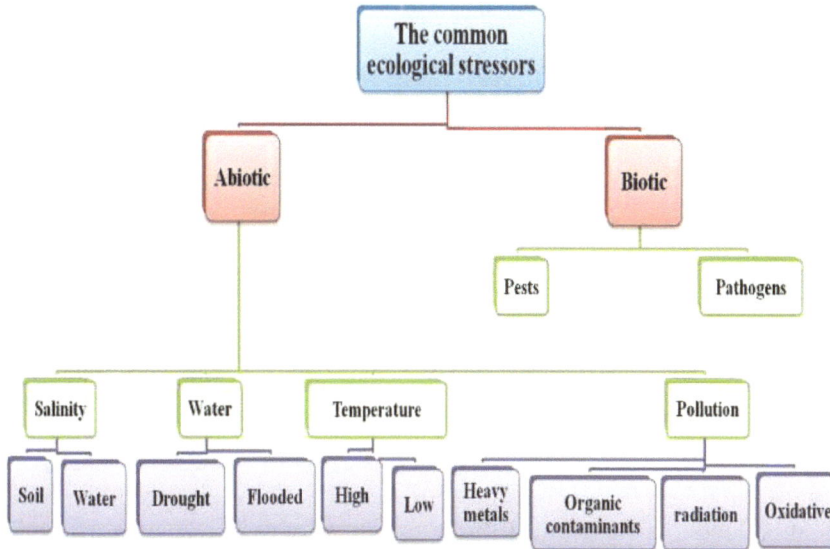

Fig. (1). The common ecological stressors that affect plant survival.

2.2. The Abiotic Stressors Under Arid And Semi-Arid Regions

Abiotic stress is the adverse effect of non-living forces on plant growth and yield below optimum levels in a specific region. The combinations of abiotic stressors are more harmful than occur solely [2].

Salinity, drought, high temperatures, and pollution with HMs are the most common stress on plants in arid and semiarid regions.

2.2.1. Salinity

Salinity is considered the major abiotic stress caused by natural events, for example, salty water, parent rock, oceanic salts, low precipitation, and higher evaporation, or by peoples' activities like incorrect cultivation management [3]. It's responsible for I) osmotic imbalance, thus low soil water availability for plants, II) ion-specific toxicities or imbalance causing decreasing nutrient uptake, consequently, affecting plant physiology (photosynthesis activity, inhibiting plants metabolism and chlorophyll content) [4]. It is estimated that about eight hundred million hectares of land are salt-affected soils [5]. The categories of saline soils are defined in soil classification literature by The Natural Resources Conservation Service (NRCS), as shown in Table (**1**).

Table 1. Classification of salt-affected soils used by NRCS [5].

Salt-affected soil Classification	pH	EC dS/m	SAR	ESP
Non	<8.5	< 4	< 13	< 15
Saline	<8.5	> 4	< 13	< 15
Sodic	>8.5	< 4	> 13	> 15
Saline – Sodic	>8.5	> 4	> 13	> 15

2.2.2. Drought

Drought is defined as the deficiency of water enough to reduce leaf water potential that reflects on the plant growth, which occurs due to a shortage in water absorption (low humidity or high salinity) concomitant to excess in transpiration (high heat or fast wind). By the year 2100, an average rise in temperature from 2–5 °C could lead to a reduction in precipitation by 15% [6]; under these conditions, the drought areas will increase.

Drought causes a decline in most crop yields; also, excessive irrigation water or the higher frequency of irrigation may result in high vegetative growth and low yield in quantity and quality [7]. Also, the lowest potato yield is produced by high water stress [8].

2.2.3. Heat

Urbanization, industrialization, and rising greenhouse gas levels are all contributing to the gradual rise in global temperature. Due to climate change-related threats, agricultural commodities will decrease [9], as well as sowing and harvesting dates will change.

All plant species' germination, development, and yield are impacted by temperature-induced stress, which is a significant environmental element. Heat has an impact on plants at all growth stages, including the tissue, cell, and even subcellular levels [10, 11]. Plant cells suffer from oxidative stress when the temperature is high. Sorghum grain yield was decreased under heat stress by decreasing antioxidant defense [12] or by increasing soil salinity.

2.2.4. Pollution

Soil pollution is a global issue, which transcends borders and whose source can be very distant. There is a need to provide proper management practices by I) long-term studies to understand the effects of continued exposure to low doses of pollutants, II) studying the synergistic and antagonistic effects between different contaminants, III) assessments and monitoring of soil pollution risks [13].

Heavy metals (Cd, Pb, As, Ni, Sb, Hg, Co, Cr, Cu, Al, Zn, Mn, and Fe) are elements with metallic characteristics and an atomic number of more than 20. These nonessential metals are considered major environmental pollutants and constitute serious hazards to human beings' health (cytotoxic and carcinogenic) and the environment. The most serious sources contributing to the contamination of soil are agricultural practices (using fertilizers, pesticides, and low-quality water), and industrial and mining activities [14 - 16]. Heavy metals toxicity depends on some soil factors including pH, cation exchange capacity, redox potential, organic matter content, and soil texture [17]. Using low-quality water for irrigation increased extractable HMs (Cd, Ni, Pb, and Cu) more than that good-quality or blended water [18].

2.2.5. The Impact of Abiotic Stressors on Plant

The negative impacts are manifested by 1) the high osmotic levels of soil solution that lead to difficulty in the water and nutrient absorption or reaches physiological drought, 2) ionic toxicity of Cl and Na that leads to nutrient imbalance. These effects negatively reflect ion absorption, transportation, photosynthesis, and metabolic processes, including the synthesis both of protein and nucleic acids. Consequently, germination, growth, and yield decrease by 20 to 50% for most crops [4, 19]. A negative relationship was observed between the increase in the mixture percentage of the drainage water with good-quality water and the nutrient contents of jatropha plants [20]. Diluted seawater water has an exception; it can increase the growth of some tolerant plants such as moringa, jojoba, jatropha, and cotton. Irrigation with diluted seawater increased moringa growth and improve its mineral content, particularly when sprayed with distilled water [21]. Wells water is naturally diluted seawater and it is the most applicable.

Drought stress leads to increasing the formation and cumulating of reactive oxygen species, osmotic pressure, and inhibiting water uptake, respiration, relative electrolyte leakage, stomatal conductance, leaf relative water content, transpiration, chlorophyll, photosynthesis, which reflects on crop growth and production inhibition [2, 22].

Heat is the major abiotic factor affecting plant morphology, anatomy, phenology, and biochemistry by inducing the rapid production of ROS that negatively influences cellular metabolic processes, protein accumulation, aggregation, and denaturation, increasing the lipids' fluidity in membranes, inhibiting mitochondria-enzymes, changes in water relations, decrease in photosynthesis, and hormonal changes [10, 23, 24].

The accumulation of HMs in plant cells may substitute for vital nutrients in enzymes or pigments disrupting their function and increasing the level of oxidative stress. Heavy metals increase ROS production and transpiration rate as well as reduce electron transport, CO_2 fixation, water potential, and sugar accumulation, resulting in inhibiting photosynthesis, which is often correlated negatively with plant growth and yield [17].

All abiotic stresses finally affect the photosynthesis and metabolism processes, as well as the transport, and storage of stress proteins and sugars, thus all techniques affect most stressors.

2.3. Ordinary Management and Rehabilitation of Soils and Plants under Stress

2.3.1. Soil Management

Understanding the limitations of the scraping technique will have positive management implications. The topsoil contains high salt concentration due to continuous evaporation as well as it contains a high level of HMs, especially in roadside soils by road traffic and wind dispersion, it can be scraped and transported out of the field. The soil electrical conductivity significantly lowered in the residual soils after scraping. Similarly, the NaCl removal rates were higher with three scrapings (97%) compared with a scraping one time [25].

One of the remediation technologies of HMs is excavation or scraping (physical removal of the contaminated material). This technique is applicable, especially in small areas that are contaminated with lead. If the new soil is not treated, pollutants will spread to another area [15, 26].

The application of organic amendments plays an important role as a biological reclamation technique of soils under stress [27, 28]. Many organic components can increase plant response to abiotic stress, such as crop residues, farmyard manure, compost, humic, ascorbic, indole acetic acid (IAA), proline, thiamine, Abscisic (ABA), kinetin, spermine, aspartic, glutathione, and ethylene [4].

Soil application of humic acid (HA) to rice plants suffered from water stress increased free proline content, decreased membrane permeability, and improved soil fertility [29].

The mulch technique considers an important practice to sustain agricultural production. Many materials have been used as mulch, like crop residue, paper pellets, gravels, rocks, litter, and plastic film. The soil surface may retain at least 30% of the plant remains, such as straw residues between rows [30]. Furrow with mulch practice decreased salinity significantly compared to furrow without mulch, this may be attributed to water saving, decreasing evaporation, raising salt leaching, decreasing salt accumulation in the rhizosphere, and increasing water intake by plants under mulch treatment [31]. The combination of earthworm and straw as mulch-treatment enhanced plant growth and nutrient concentration compared with the individual addition of earthworm or straw [26].

Vermicompost is associated with the encouragement of tomato growth, yield quantity, quality, and improvement of the chemical, hydro-physical and biological soil characteristics under abiotic stress (salinity and water deficit) as concluded by García *et al.* [29]. The highest dry weight of bean, 100 seed weight, seed yield, and seed content of N, P, and K were obtained with 100% of nitrogen recommendation as biogas residues compared with ammonium nitrate under water deficit conditions (75% of WR) [32].

Compost burning during the seasons of cultivation increased the number of effective groups blocked with Cd ions [14, 33]. Enriched banana compost with effective microorganisms had a promising effect on reducing Cd, Ni, and Pb concentration in soil and plant compared with mineral fertilizers [33].

Soil Microorganisms such as fungi (arbuscular mycorrhizal), yeast, azola, and rhizospheric microbes (Pseudomonas, Azotobacter, Azospirillum, Rhizobium, Pantoea, Trichoderma, Bacillus, Enterobacter, Burkholderia, Bradyrhizobium, Methylobacterium, and Cyanobacteria) contribute to alleviating abiotic stresses in the plants by their metabolic and genetic capabilities. Soil inoculations with microorganisms improve plant growth, dry weight, and yield under abiotic conditions, This increment may be attributed to 1) modifying root elongation, shoot length, leaves anatomy (thickness of the blade, the vascular bundles' dimensions, and the number of xylem vessels), 2) increasing minerals availability, nutrient uptake, K/Na ratio, soluble sugars, osmosolutes, photosynthetic pigments, and protein production even in nutrient-poor soils, 3) reducing proline concentration and oxidative damage, 4) keeping higher antioxidant enzymatic activities, 5) increasing water absorption capacity, the efficiency of photosystem and stomatal conductance [34 - 36]. Although microorganisms can't degrade

HMs, they can influence their movement and transformation by altering their chemical and physical characteristics [26].

2.3.2. Water Management

Salts are most efficiently leached under short irrigation intervals (higher frequency of irrigation). Keeping adequate soil moisture content between irrigation times effectively dilutes salt concentrations in the rhizosphere, thus reducing the salinity hazard. The low frequency (the wide period between irrigations time) induced water stress, in contrast to the high-frequency treatment, the intensity of this influence depends on soil-physical properties [37, 38].

All efforts must to synergistic to rise the irrigation water use efficiency (IWUE) of economic crops to confront the shortage in freshwater. IWUE = Grain, oil, or any economic yield (kg/fed.) /total water applied (m^3/fed.) [8, 32]. One cubic meter of irrigation water can produce 1.7 kg grains, 2.0 kg ears, and 3.4 kg biological yield of maize [38]. High variation among IWUE values occurs by the difference in irrigation systems [39] and irrigation water deficit [38]. Irrigation of maize plants with a subsurface drip irrigation system gave a higher IWUE value than surface one [37]. Usually, the application of the highest irrigation rate produces the smallest values of IWUE and EWP [8, 31, 40]. IWUE values for maize were 0.94, 1.14, and 1.5 when 3202, 2722, and 2241 (m^3/fed.) were applied, respectively [38]. Compost application increased IWUE values of maize and wheat more than mineral fertilization treatment under the rain and the addition of supplemental irrigation (SI) as shown in (Fig. **2**).

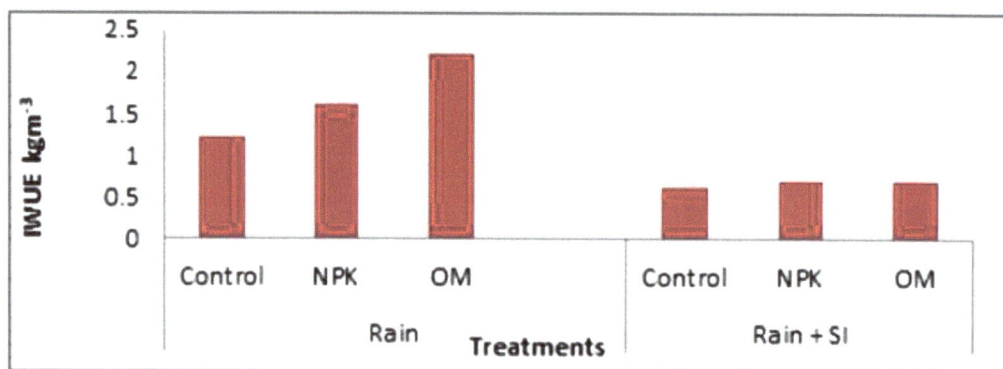

Fig. (2). Effect of NPK and organic matter fertilization on IWUE values of maize crop under the rain and supplemental irrigation [41].

The direct aim of a reshaping of field drains is to lower the water table level, and optimize soil-physical properties, thus enhancing the removal of salts by the leaching process [4]. The chemical leaching is cleaning the polluted soil by using water and some chemical reagents such as phosphoric acid, potassium phosphate, EDTA, citric acid, citric acid + potassium phosphate, and saponin tea, they can effectively remove HMs [26]. Using EDTA and ferric chloride in soil washing technology could result in an ecologically safe, feasible soil heap leaching technique of HMs remediation (Cd, Cu, and Zn) from various soil types [16].

Soils are destroyed by excessive sodium that dispersed clay mineral particles. Calcium has an effective role in replacing Na^+ in salt-affected soils. Significant positive influences of applied calcium were observed on leaf number, leaf area and plant dry weight, photosynthetic pigments, nutrient uptake, and cotton yield compared with the control (without calcium addition) as described by Dewdar and Rady [42]. Increasing calcium carbonate content in soils, particularly the fine form significantly increased the adsorption of Cd, Ni, and Pb as an active portion in the process of adsorption [43]. The addition of phosphorus to the contaminated soil decreases the harmful effect of HMs by creating less toxic compounds (heavy metal phosphates) [15]. The application of this technique required a high-quality artificial drainage system in addition to the drainage water should be treated.

Magnetic or electro-magnetic treated water is considered one of the promising techniques for improving growth. Fig (3) illustrates one of the electromagnetic apparatus and its pipe before and after use. Irrigation with magnetized water increased the growth parameters of cowpea and raised stomatal conductance, IWUE. The influence of magnetic water may be attributed to the increase in root growth and stomatal conductance that increase the absorption of nutrients [44]. Also, available soil N, P and K were significantly increased by using different magnetized water sources [45], particularly under saline conditions [46].

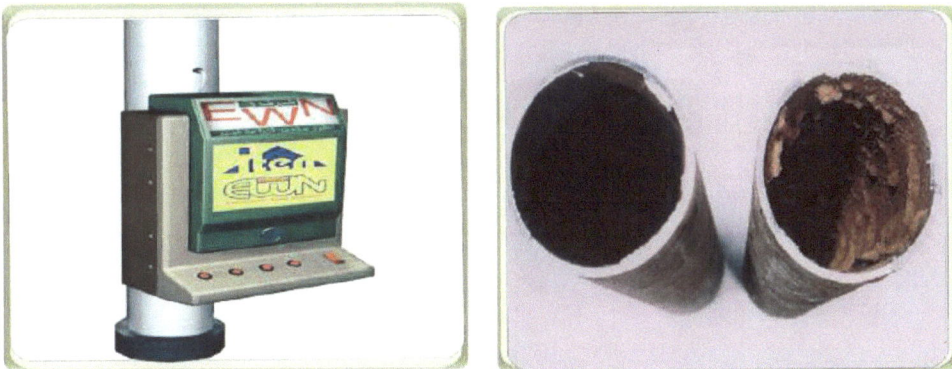

Fig. (3). One of the electromagnetic apparatus and its pipe before and after use.

2.3.3. Crop Management

Beyond the roles of soil scraping, and chemical leaching, electrical, and incineration techniques for heavy metal purification are costly and may degrade the soil. Plant breeding (selection of crop genotypes that can accumulate salts and HMs) and phytoremediation technology has been applied as promising, environment-friendly, and low-costly methods [18, 47, 48].

Halophytes and phytoremediation plants are a group of plants able to tolerate salinity and absorb high concentrations of HMs. It has medicinal and ecological values; protects the shores, provides wood and oil for fuel, fodder, and shelter to animals,...*etc. Kochia indica* and *Artemisia monosperma* are native wild, forage, salt-tolerant and grown well in coastal salt marshes in Egypt Fig (**4**), the former used as an anthelmintic [49, 50]. Some halophytes species can absorb salts and HMs in both saline and contaminated soils [51].

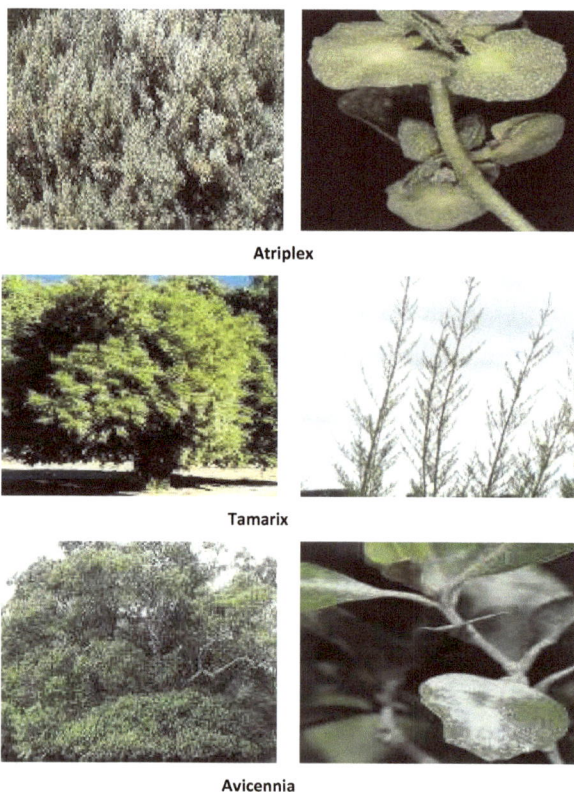

Atriplex

Tamarix

Avicennia

Fig. (4). Atriplex, Tamarix and Avicennia trees.

In arid zones, Atriplex and Salicornia species are the dominant saltbushes that tolerate abiotic stress (drought, heat, salinity, and pollution) utilized as cattle feed and used in revegetation of soil to prevent desertification, as well as employed in phytoremediation of contaminated soils [52 - 54]. Canola plant [47] and flax varieties, especially Sakha 1 are high tolerance to irrigation with drainage and blended water, this may be due to HMs excluded, retained at the root level (Its content is higher in the straw than that in seeds), or transformed into physiologically tolerant forms [17, 18].

Appropriateness of balanced nutrient availability is important to maintain soil fertility, and plant productivity, and to reduce environmental pollution [13]. The balanced fertilization plays a crucial role in the plant life cycle begins with seed production and continues through plant growth and protection. The sufficient nutrient supply leads to withstand biotic and abiotic stresses, making it cost-effective and advantageous under stressful circumstances [19, 55].

Effective nutrient management (N, P, K, Ca, Mg, Zn, and Se) plays an important role to enhance plant resistance to heat stress by 1) helping in stomatal regulation, 2) helping in osmotic adjustment, 3) assisting in preserving high water potential in plant tissue, 4) activating the physiological, biochemical and metabolic processes [10, 24]. Selenium (Se) and Salicylic acid (SA) addition can improve the plant tolerance to temperature stress by raising the activity of antioxidant enzymes and lowering membrane damage induced by ROS [10].

The balanced NPK fertilization is essential for improving growth, nutrient content, and mineral ratios in jatropha plants that are irrigated with low-quality water and enhancing their ability to survive with abiotic stress [20]. Zinc sulfate has a number of significant advantages, such as enriching crop micronutrients, enhancing cereals' ability to withstand salt stress and reducing the uptake of toxic elements [30]. Application of potassium (KNO_3) surpasses Zn-EDTA and ascorbic acid treatments, particularly under high salinity levels [50]. Boric acid application mitigates the salinity effect on germination percentage, it makes water and food transport easier by reducing stomata number, the distance between vascular bundles, transpiration and water loss [56]. The foliar application of diluted phosphoric acid increases root growth and promote the ability of plants to absorb nutrients, especially under salinity stress condition [57].

Macronutrients plus micronutrients in addition to Si, Se, Na and V could protect crop plants against abiotic stress by 1) regulation of metabolic activities and synthesis of biomolecules, 2) activation of the antioxidant defense system, thus improving the plant resistance power, 3) maintenance of ionic homeostasis, 4) Nutrients are essential for stomatal conduct, osmoregulation, enzyme activity, and cell growth. 5) energy reactions in metabolism, protein- and nucleic acid-synthesis processes, 6) the movement of electrons during photosynthesis and the production of riboflavin, beta-carotene, and ascorbic acid; 7) auxin creation and root improvement [55, 58, 59]. The application of Ca can play an important role to enhance plant tolerance against HMs, particularly Cd toxicity, because calcium is an essential nutrient for plant growth, osmotic adjustment, membrane integrity, and signaling transduction [54].

The foliar application of growth regulators as ascorbic acid (400 mg/l) at 40, 60, and 80 days after sowing induces abiotic stress tolerance (salinity up to 7000 ppm) of cotton plants, thus leading to favorable growth and a high economic yield [42]. Also, lead to an increased stem, leaves, and seed protein of jojoba plants grown under stress, compared with that treated with distilled water [60], because it plays a main function in cell division and differentiation thus modifying the physiological and morphological characteristic of plants [50]. The positive effect of growth regulators in mitigating abiotic stress may be attributed to 1) increasing stress-protective protein production, 2) improving mineral nutrition, by raising calcium accumulation as well as decreasing sodium and chloride absorption, 3) avoiding cellular damage provoked by oxidative stress by scavenging of hydroxyl radicals, and 4) Regulating the water balance of plant tissues [4, 49].

2.4. Modern Techniques to Combate Abiotic Stress

2.4.1. Nano-technology

Nanotechnology is a modern issue adopted to maintain long-term sustainability and protect crops, which plays on the meaning of "maximum effect from the low dose". Nano-agrochemicals and clay-based nanomaterials are characterized by a controlled release that increases the efficiency and stability of the used material. Nano form of some elements oxide such as iron, zinc, copper, and magnesium has promising growth-promoting and antimicrobial activity against phytopathogens. Usability of nano-based materials also applied in plant nutrition, infection detection, plant protection, gene delivery, soil, and water nano-remediation and measurement of pesticide and nutrient levels, *etc.*, [22, 55, 61 - 63].

In addition, nanotechnology could be used to detect and treat plant diseases at the molecular level, as well as improve plant nutrient absorption and manage salt-affected soils. The nanoform of zeolite, calcium, and biochar can bind sodium ions on exchange sites thus decrease of clay dispersion [64, 65].

Nanoparticles of micronutrients are being significantly used as a foliar application, which is cost-effective [66]. Nano-Si and nano-Se have great efficiency in mitigating the adverse effect of abiotic stress, especially the pollution of HMs, thus raising the yield of several crops under stress [67]. The application of nano-silica is beneficial in enhancing the germination and growth of 25 lentil genotypes and its addition could stimulate the defense mechanisms against salinity stress consequently improving salt tolerance [68]. Nano-silica (7.0 nm) is also used to enhance sweet pepper growth under salinity stress by enhancing turgor pressure, water translocation and xylem humidity consequently, leaves relative water content and IWUE will be improved [55]. This advantage may be due to 1) the very fine particles of nanomaterial have a high specific area (1 g of 7.0 nm nano-silica shows absorption surface equal to 400 m^2), thus increasing the solubility of some low-soluble materials, 2) nanomaterials can encourage the roots regeneration [55, 67, 69]. Nano-silica promotes plant growth and production and can be employed as nanopesticides, nanoherbicides, and nanofertilizers. Also, they are boosting plant resilience to biotic and abiotic challenges [70]. Some of the nanomaterials' mechanisms that enhance stress tolerance in plants are (1) improving the plant's ability to exclude sodium, (2) enhancing the plant's ability to retain potassium, (3) maintaining reactive oxygen species (ROS) homeostasis, (4) enhancing the production of nitric oxide, (5) reducing lipoxygenase activities to decrease membrane oxidative leakage. (6) increasing the content of soluble sugar by improving α-amylase activities, and (7) reducing the length of root apoplastic barriers to transport Na^+ to shoot [71], (8) regulating the antioxidant defense system, (9) modulating phytohormones [72]. (10) boosting the expression of genes involved in defense [70].

On the other hand, several investigators recorded that both forms of an element (ordinary particle size and nano-size) gave the same effect; moreover, the highest dose of nanomaterial produces a negative effect. Haghighi and Pessarakli [73] reported that both Si and nano-Si treatments enhanced the mesophyll conductance and photosynthetic rate of tomato plants as well as its WUE under salinity stress conditions. In another study on tomatoes grown under high-temperature stress, both selenate (ordinary source of Se) and nano-Se had a similar beneficial effect on some growth metrics, at low concentration. In addition, Se-selenate was superior to nano-Se in enhancing shoot and root biomass [74]. Carbon nanotubes at a rate of 0.05 mg/l increased the number of germinated embryos, root number, and root length of date palm tissue culture, but at a rate of 0.1 mg/l, the growth and nutrient concentration were decreased [75]. Although the detrimental effects of salinity on cotton plants were reduced by the foliar application of 200 ppm nano-Zn, the P/Zn ratio was impacted negatively; thus, phosphorus fertilizer should be applied in addition to nano-Zn treatment to prevent P/Zn imbalance [76]. Fertilizer treatments (SiO_2, ZnO, SiO_2+ZnO, nano SiO_2, nano ZnO, nano SiO_2+nano ZnO, and control) did not affect any measured trials except for ZnO which produced 12.5% increase in cotton lint yield than control [77]. Different amounts of nanoparticles can have both beneficial and harmful biological effects [78]. The spraying of iron EDTA increased the plant height, and the fresh and dry weight of corn compared with nano-Fe, this result may be attributed to the defensive response of the plant towards the nano-iron as indicated by the appearance of four silicon units arranged like petals of a flower in the SEM micrographs of nano Fe-sprayed leaves as illustrated in (Fig. **5**) [79].

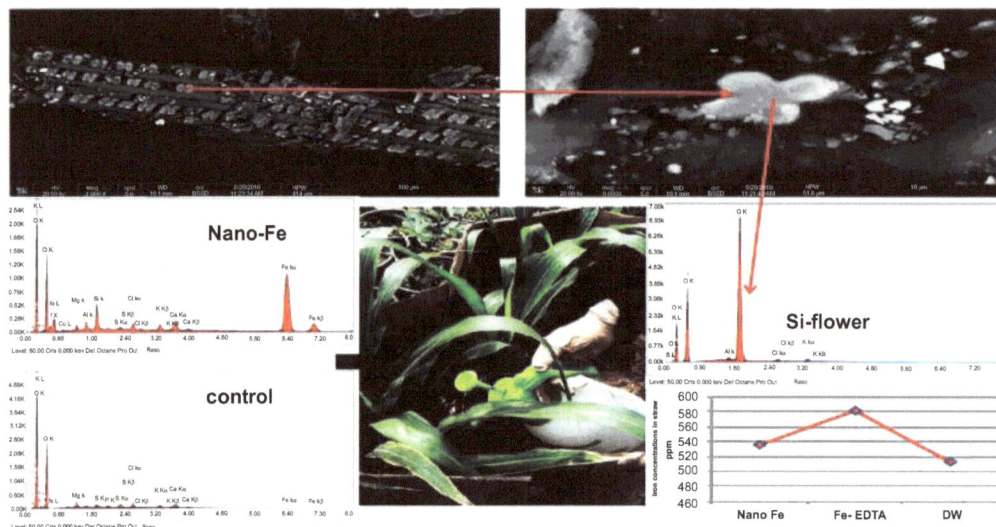

Fig. (5). Comparison between iron EDTA and nano iron.

2.4.2. Intelligent-green Composites

It is necessary to develop new applications which are sustainable agronomically, economically, and environmentally to produce safe food for human consumption. Many efforts should be synergistic to finding modern ways to reduce the application of mineral fertilizers, because they are the main reason for soil degradation (salinity and chemical pollution), and replace them with environmental friendly fertilizers without affecting productivity. Thus, there is a need to produce some Intelligent-green composites that can increase crop production under stress conditions. Several products achieved to be called Intelligent-green composite such as agriglass, organo-clay, bio-clay, bio-straw, polymers, Silicate, salicylic and selenium mixtures, kinetin + spermine, and ABA + Si compositions,...*etc.*

Glasses network (such as phosphosilicate glass) has an exclusive ability to include different chemical compounds into their structure [80] and release them gradually [81, 82]. The addition of agriglasse (glassy slow-release fertilizer) has a significant ecological effect and produced higher records than that of ordinary mineral fertilizers in the measurements of plant height, ears and cob weight [81, 83]. Ears, grains, straw and biological maize yield increased with increasing agriglass rates under water stress conditions. Although, under normal conditions of irrigation (100% of water requirement) the coarse size of agriglass granules (1mm) is superior to the fine one (0.5mm), under water stress (85 and 70% of water requirement) the fine diameter is the best [82].

Natural materials, including clay minerals and zeolites groups, play a vital function in decreasing the potential hazards of toxic substances. The arrangement of the clay minerals as their ability to adsorb lead followed this order: zeolite > montmorillonite > bentonite. This study suggests that using available natural materials could be an economic and promising alternative solution in contaminated water to minimize the hazards of HMs. The clay minerals are suitable materials for heavy metal removal from industrial wastewater [84]. This technique is called stabilization or in *situ* fixation of HMs. The viability of using organo-clay complexes is interesting for minimizing the impact of both heavy metal and organic pollutants [85]. Organo-clay complex revealed a high ability as a sorbent to Cd, Cu [86], Pb, and Hg from contaminated water [87]. A new lignocellulose/montmorillonite nanocomposite was prepared by chemical intercalation and functioned as an effective adsorbent of Ni from wastewater [88]. In the absence or presence of compost, clay minerals can increase plant development in polluted environments and lower the content of HMs [89]. Alum–polyaluminium chloride coagulation aided by organically modified nano-montmorillonite as an adsorbent material was superior to nano-bentonite or nano-halloysite for the separation of multi-pesticides from low-quality water [90]. The intercalation of Quillaja saponin onto montmorillonite clay mineral (organo-caly) significantly improves the performance of the clay in the removal of the organic contaminants in wastewaters [91]. Clay minerals could be considered an effective supporting input as a microbial carrier [92]. The maximum values of NPK-use efficiency, IWUE, and EWP of bean yield were associated with the interactions between biozeolite (intercalation of biofertilizer onto zeolite clay mineral) with a high rate (450 Kg/fed.) with irrigation by both rates of 100 and 85% of bean water requirements [93]. Biozeolite proved its ability to reduce the impact of salinity and nematode on bean plants Fig (**6**).

The percentages of net mortality of *M. incognita* juveniles were 97% in biozeolite treatment compared with 37% in sole zeolite application [92]. Also, straw cold is considered an important microbial carrier (bio-straw), the combination of azolla with cyanobacteria increased the soil organic matter OM and N, P, K (available in soil and plant content), slightly decreased soil EC and demonstrated various advantages over chemical fertilizers and increased saline soil fertility [94].

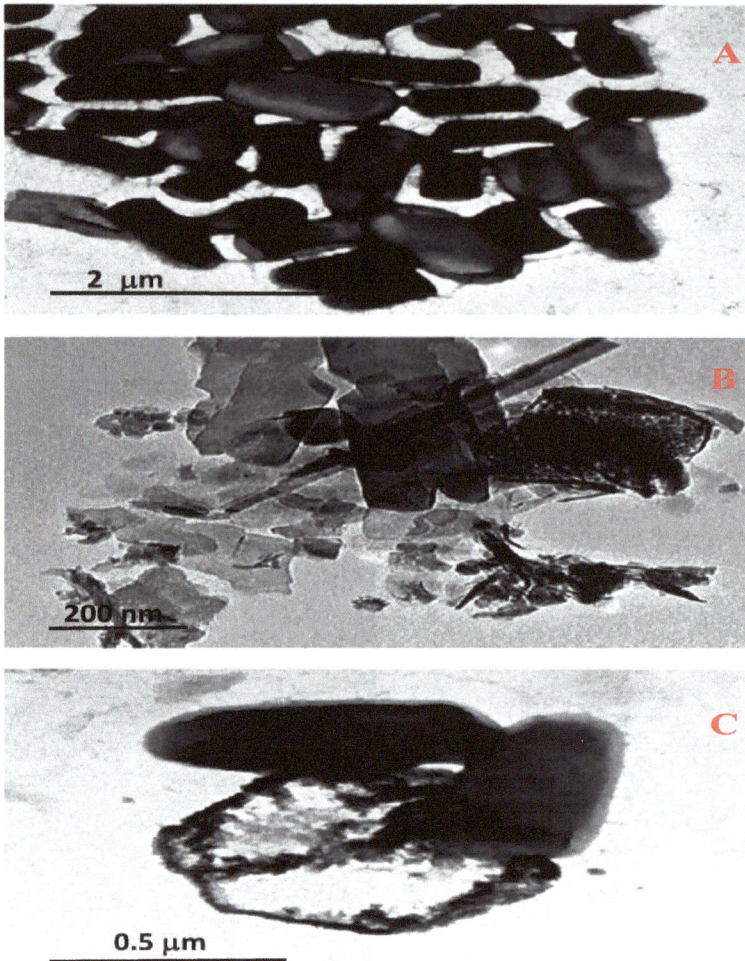

Fig. (6). TEM of rhizobium (A), zeolite (B) and (C) biozeolite.

There are new polymers that provide high-capacity chelating agents for alleviating ecological stress. Both dendrimers and hyperbranched have numerous polar functional end groups which react strongly with heavy metals to be in a form that is much less harmful [95]. The hyperbranched polyesters (carboxyl or hydroxyl-terminated polymer) exhibit great potential as sorbents for removing HMs. Carboxyl terminated polymer was superior to those of hydroxyl-terminated ones, and the selectivity follows the rank: Cu > Cd > Pb for all polymers, irrespective of the kind of terminal group [96]. Modified nano-polysaccharides are highly promising biosorbents for capturing heavy metal ions (Ni, Cr, Zn, Ag, Cu, and Fe) from wastewater and may enable next-generation water purification technologies [97].

The usage of the hydrophilic polymer in soils, particularly in light-structure soils, increases the hydrophysical properties of soil, like water holding capacity and moisture content, as well as improves crop yield and water use efficiency [98]. Hydrogel (HG) is a group of polymers that can absorb high quantities of water without dissolving, and they vary in their origin and composition. Although HG wasn't fully efficient under saline conditions, its application significantly improved the growth parameters of some vegetables and decreased the electrolyte leakage of plants and the electrical conductivity of soil [99]. Soil water content in the treated soil with HG increases to about 2.3 times more than in the untreated one. Application of HG can increase the period between irrigations and IWUE in coarse-textured soils, this is a vital approach in arid and semi-arid regions. The addition of hydrogel increases seedling growth (2.7 times), elongates the roots (3.5 fold), enhances calcium uptake, improves ion relationships of salt-tolerant woody species, *Populus euphratica*, and its ability to exclude salts more than those grown in untreated soil [100].

Although straw-based hydrogel Fig (7) decreased ear yield by only 4% less than applying acrylamide hydrogel, the former which was based on rice straw as an agricultural residue, is the lowest price and free of acrylamide (environmentally friendly product). Generally, the application of HG improves the maize yield, minerals concentration, and uptake, as well as both IWUE and nutrients, use efficiency compared with control [38].

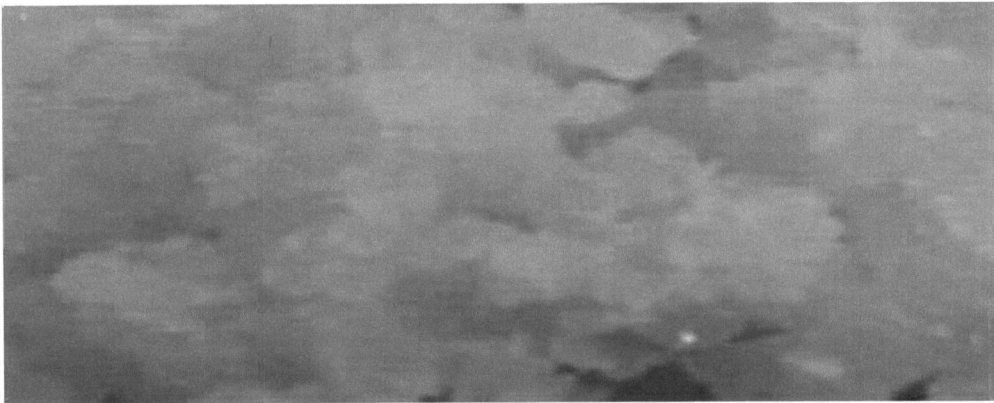

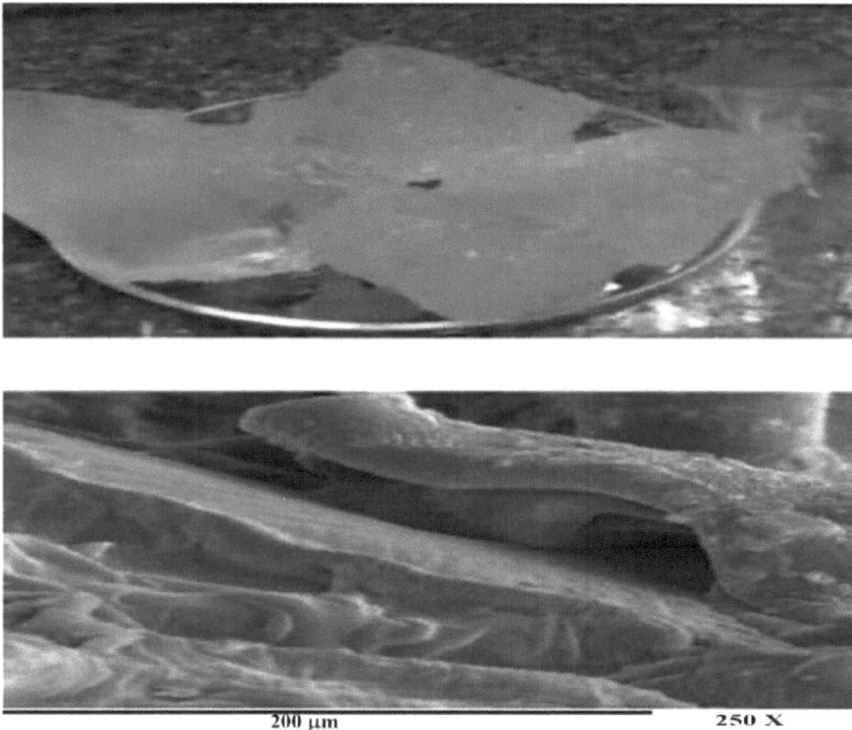

Fig. (7). Straw-based hydrogel after swelling in distilled water.

Silicon (Si) is the second most common element in the crust of the Earth. Although Si has all conditions that should be realized in the nutrient elements (I-Si is a fundamental component of the plant structure and metabolism, II- The reduction of Si in plant growth media shows deficiency symptoms at various development phases of plants, III-The supply of Si to plant growth media can manage the deficiency symptoms), it has not yet been recorded among the essential nutrients for plants. Silicon is known for effectively mitigating material to different biotic and abiotic stresses such as nutrient imbalance, drought, salinity, heat and metal toxicity, resistance to pests and diseases, *etc.* [101]. Potassium silicate application significantly increased basic branch, plant height, bod number, stover and biological yield under salt stress conditions [40]. Salinity impact on Murcott trees could be mitigated by treatments with potassium silicate. Fruit yield, its quality and the root distribution, as well as photosynthetic pigments and mineral status of leaves, were enhanced by spraying application of potassium silicate [102]. The mechanisms that describe the importance of Si in mitigating abiotic stress are concluded in [23, 31, 101, 102]: 1) improving leaf erectness, chloroplasts structure, metabolism, photosynthetic activity, 2) reducing Na and Cl uptake and their translocation in plant parts 3) enhancing K content, K/Na ratio, chlorophyll content, 4) Si deposits in cell walls of different plant parts and modified the cell wall architecture, 5) silica deposition in leaves decreases the permeability of the plasma membrane, protects cell membrane of damage, limits transpiration and improves plant water balance and its storage in plant tissues, 6) Si is reducing ROS creation, 7) stimulates of antioxidant systems in plants, and 8) Si play as a co-precipitation agent of the toxic metal ions. From the above-mentioned review, it can be confirmed that silicon is one of the necessary nutrients to plant growth, development, and protection.

Salicylic acid (SA) is a bioregulator of phenolic nature and is considered a hormone. The salts of salicylic acid are called salicylates. It can ameliorate the negative effect of abiotic stress by 1) its preventive role of cell membranes that raise the plant tolerance to damage, 2) increasing division and elongation of the cells, 3) the osmoregulation effect that is possibly mediated by increased production of sugar as well as proline. Application of SA (200 ppm) to cotton plants recorded the highest increment in growth parameters, nutrient uptake, Ca:(K+Na) ratio, and raising its alleviation of salt stress [103].

The mixture of potassium silicate and salicylic regulates moringa's response to salt stress and enhances its growth and mineral status [21].

Selenium (Se) is an essential element of the human body and an antioxidant, that prevents exposure to cancer. Application of Se and Si in combination or solely have been used in mitigating the adverse effect of abiotic stress, particularly HMs such as lead [104] and cadmium [105]. The highest grain production was obtained when 75 mg L^{-1} of selenate (a source of Se) was applied to plants cultivated under high-temperature stress [51]. The addition of selenium to a wheat plant grown under stress conditions enhances transpiration rate and photosynthetic, and chlorophyll contents, thus increasing the growth. This increment may be due to 1) reducing membrane injury by increasing antioxidant enzyme activity, 2) increasing osmoprotectants like soluble sugar, proline, and soluble protein 3) reducing ROS accumulation [51, 106].

The beneficial effects of Si and Se are enhancing plant-nutrient balance, plant growth, development, and yield, increasing resistance to parasitism and herbivores, mitigating biotic (plant diseases) and abiotic stress (drought, temperature, salinity, water, toxicity, and oxidative stress) [55].

A synergistic effect was found among some materials such as (Silicate, salicylic, and selenium), so their mixtures produced highly effective growth regulators.

The protective influence of the (spermine + kinetin) composite on *Vigna Sinensis* plants under water stress conditions appeared mostly due to increased chlorophylls and protein content [107]. Application of (ABA+Si) composite led to ameliorating rice growth *via* lowering Na^+/K^+ ratios, thus increasing the net assimilation rate and stomatal conductance of seedlings [108].

2.4.3. Genetic Engineering

The utilization of omics techniques (Studies on the genome) has contributed to the development of plant tolerance to abiotic stress. Ongoing advances in the branch of biotechnology and transgenic techniques, alongside the integration between screening and breeding, make it possible to produce transgenic genotypes to perform better under environmental stress conditions [34].

The mechanisms of genetic control of stress tolerance in plants are still poorly understood because of their complexity. Abiotic stress tolerance is regulated by a variety of genes in different species. Consequently, genetic diversity can only be indirectly shown by observing how different genotypes respond to different stimuli. Many proteins found in plants seem to be involved in the abiotic stress response. These proteins may be able to modulate ion homeostasis as well as the levels of osmolytes (mannitol, fructans, proline, and glycine betaine). Although transgenic and molecular breeding is more costly than common conventional breeding, it provides an efficient way to create salt-tolerant cultivars [30].

It will be necessary to create salt-tolerant cultivars that are better able to handle the escalating salinity stress using both standard breeding methods and genetic modification. In Egypt, numerous researches are carried out to increase the tolerance of tomato and wheat to osmotic stress by transferring osmoregulatory genes from barley and yeast [49]. Carrera *et al.* [109] reported that four of the top twelve gene subnetworks were related to biotic and abiotic stresses. Plant-specific NAC transcription factors play an important function in plant resistance to abiotic stress [110]. Rice responds better to salt and drought stress when the OsNAC2, OsNAC6, and OsNAC045 genes are overexpressed [111]. Also, WRKY is a group of proteins that have a highly conserved WRKYGQK domain; it responds to biotic and abiotic stresses. Overexpression of WRKY25 or WRKY33 genes can improve the salt tolerance of *A. thaliana* [110].

2.5. Economic Aspects

Arid and semi-arid regions are characterized by high population and lower per capita income than moderate living standards. Crops are exposed to biotic and abiotic stressors such as salinity, heat, toxic agrochemicals, pathogens, pests,…*etc.*, which affect human health and the economy. In irrigated soils, the global annual cost of the degradation in salt-affected soils could be US$ 27.3 billion by the depression of crop yield [112]. One of the common stressors that lead to the country's economic insecurity is drought [22].

Choosing salt-tolerant crops and managing salt-affected soils are vital strategies to promote the agricultural economy [3, 63]. In 2000, about AU$ 1.4 billion as investments over seven years were given by Australia's National Action Plan to support actions by land managers and communities in salt-affected soils, because the cost of engineering and management of saline soils (scraping, irrigation, and agricultural drainage systems) is very high [30].

Using crop residues such as rice and sugarcane as sources of Si for combating environmental stress is cheaper than chemical compounds like sodium or potassium silicate [38]. Application of plant growth-promoting bacteria, antioxidants, and nutrient management techniques might allow for low-cost crop production under stress conditions [34].

There are three main techniques to clean up the contaminated soils: excavation (removing the contaminated soil), in *situ* fixation (adding different materials to decrease the availability of contaminants), and phytoremediation. Economically, the first technique is the most expensive method when large amounts of soil must be transferred as toxic waste is required. The second technique may be about half the cost of the first one and both of them take about the same amount of time. The third technique is the slowest method but has the advantage of being a very cheap, natural solution to the environmental pollution issue [15].

3. A FUTURE VISION/ CONCLUSION

To overcome the harmful effects of abiotic stress on the crops, one or more of the following strategies can be applied: 1) scraping and removal of saline or contaminated top soil, in addition to the replaced soil should be treated or entered in building materials to be stabilized, 2) land leveling, deep ploughing adapting irrigation and drainage systems, 3) chemical leaching of salts or other contaminant with adding calcium, phosphorus, EDTA,...*etc.*, 4) usage of mulch technique, 5) using halophyets and phytoremediaton plants, 6) balanced fertilization, 7) addition of organic matter, especially compost and biogas residues, 8) Usage of promoting growth solutions (humic, ascorbic, amino acids, phytohormones, phosphoric, salicylic and silicate, *etc.*), 9) use of polymers as a soil conditioner and remediators, 10) incorporating of halophilic microorganisms carried on clay minerals, 11) Application of some green composites such as straw based hydrogel, straw based agriglass, organoclay, bioclay, biostraw, kinetin + spermine, ABA + Si and Si+SA+Se, 12) Usage of nano clay, nano carbon and nano zinc with some restrictions, and 13) selecting the tolerance plants that produced by breeding and genetic engineering techniques.

There is an obvious need for more studies on some new research points as follows:

- Future studies are required for a better understanding of the status of global soil pollution, especially with transferring pollutants from the source to the customer in agricultural products form (fruits, vegetables, *etc.*,). Thus, the integrated recommendation of polluted soil reclamation should be documented.
- Further studies on using of clay minerals as a carrier for microorganisms to produce intelligent compounds that work as biofertilizers and biopesticides under the interaction between biotic and abiotic stress are required.
- Further studies are warranted to explore different concentrations of silicate, salicylic, and selenium to arrive at the appropriate mixing percentages.
- Electrical and magnetic mechanisms in treating saline and low-quality water are not fully understood, thus further studies in this field are required.
- After more than 20 years of nanotechnology scientific research, applications of nanomaterials are still between acceptance and rejection in the soil fertility and plant nutrition field. The tissue culture technique could be the most suitable method for studying the impact of nanotechnology on the plant. Furthermore, nano-cellulosic is one of the cheapest organonanomatrials, and it has demonstrated potential applications in a wide array of manufacturing sectors, such as nano-silica products. In many studies, the second application rate of different kinds of nanomaterials reverses the positive result and decreases plant growth compared with the first dose. To date, further studies are needed into the potential of nanomaterials in boosting plant tolerance to abiotic stress, the accurate molecular mechanisms, that elicit signaling responses in plant cells, and the impact of nano-fertilizers on the environment and users' health.

Finally, everyone must make sure that all their activities do not increase environmental stress.

REFERENCES

[1]   Van Velthuizen H. Mapping biophysical factors that influence agricultural production and rural vulnerability. Food & Agriculture Org 2007.

[2]   Cramer GR, Urano K, Delrot S, Pezzotti M, Shinozaki K. Effects of abiotic stress on plants: a systems biology perspective. BMC Plant Biol 2011; 11(1): 163.
[http://dx.doi.org/10.1186/1471-2229-11-163] [PMID: 22094046]

[3]   Rengasamy P. Soil processes affecting crop production in salt-affected soils. Funct Plant Biol 2010; 37(7): 613-20.
[http://dx.doi.org/10.1071/FP09249]

[4]   Abou-Baker NH, El-Dardiry EA. Integrated management of salt affected soils in agriculture: Incorporation of soil salinity control methods. Academic Press; 2015.

[5]   www.fao.org/ag/agl/agll/spush2008.

[6]   Ciscar JC. The impacts of climate change in Europe (the PESETA research project). Clim Change 2012; 112(1): 1-6.
[http://dx.doi.org/10.1007/s10584-011-0336-x]

[7]   Gaur PM, Tripathi S, Gowda CL, *et al.* Chickpea seed production manual, Inter. Hyderabad: Crops Res. Institute for the Semi-Arid Tropics 2010; p. 28.

[8]   Eid TA, Ali SM, Abou-Baker NH. Influence of soil moisture depletion on water requirements, yield and mineral contents of potato. J Appl Sci Res 2013; 9(3): 1457-66.

[9]   Shakoor U, Saboor A, Ali I, Mohsin AQ. Impact of climate change on agriculture: empirical evidence from arid region. Pak J Agric Sci 2011; 48(4): 327-33.

[10]   Waraich EA, Ahmad R, Halim A, Aziz T. Alleviation of temperature stress by nutrient management in crop plants: a review. J Soil Sci Plant Nutr 2012; 12(2): 221-44.
[http://dx.doi.org/10.4067/S0718-95162012000200003]

[11]   Bac-Molenaar JA, Fradin EF, Becker FFM, *et al.* Genome-wide association mapping of fertility reduction upon heat stress reveals developmental stage-specific QTLs in Arabidopsis thaliana. Plant Cell 2015; 27(7): 1857-74.
[http://dx.doi.org/10.1105/tpc.15.00248] [PMID: 26163573]

[12]   Djanaguiraman M, Prasad PVV, Seppanen M. Selenium protects sorghum leaves from oxidative damage under high temperature stress by enhancing antioxidant defense system. Plant Physiol Biochem 2010; 48(12): 999-1007.
[http://dx.doi.org/10.1016/j.plaphy.2010.09.009] [PMID: 20951054]

[13]   Be FAO. Be The Solution to Soil Pollution, Outcome Document of The Global Symposium on Soil Pollution. Rome, Italy, 2018; 1-32.

[14]   Joshi UN, Luthra YP. An overview of heavy metals: impact and remediation. Curr Sci 2000; 78(7): 773-4.

[15]   Lambert M, Leven BA, Green RM. New methods of cleaning up heavy metal in soils and water. Environmental science and technology briefs for citizens 2000; 7(4): 133-63.

[16]   Bilgin M, Tulun S. Removal of heavy metals (Cu, Cd and Zn) from contaminated soils using EDTA and FeCl3. Glob NEST J 2016; 18(1): 98-107.
[http://dx.doi.org/10.30955/gnj.001732]

[17]   Shah FU, Ahmad N, Masood KR, Peralta-Videa JR. Heavy metal toxicity in plants. In Plant adaptation and phytoremediation. Dordrecht: Springer 2010; pp. 71-97.
[http://dx.doi.org/10.1007/978-90-481-9370-7_4]

[18]   Atwa AAE, Abou-Baker NH, Nsaar MMI. Soil properties, flax varieties production and their heavy metals content under irrigation with low quality water. J Soil Sci Agric Eng 2014; 5(5): 675-85.
[http://dx.doi.org/10.21608/jssae.2014.49349]

[19]   Noreen S, Fatima Z, Ahmad S, Athar HU, Ashraf M. Foliar application of micronutrients in mitigating abiotic stress in crop plants In Plant nutrients and abiotic stress tolerance. Singapore: Springer 2018; pp. 95-117.
[http://dx.doi.org/10.1007/978-981-10-9044-8_3]

[20]   El-Ashry SM, Hussein MM, Nesreen H. The potential for irrigating jatropha with industrial drainage water under mineral fertilization. J Genet Environ Resour Conserv 2013; 1(3): 276-86.

[21]   Hussein M, Abou-Baker NH. Growth and mineral status of moringa plants as affected by silicate and salicylic acid under salt stress. Int J Plant Soil Sci 2014; 3(2): 163-77.

[http://dx.doi.org/10.9734/IJPSS/2014/6105]

[22]   Kumar P, Rouphael Y, Cardarelli M, Colla G. Vegetable grafting as a tool to improve drought resistance and water use efficiency. Front Plant Sci 2017; 8: 1130.
[http://dx.doi.org/10.3389/fpls.2017.01130] [PMID: 28713405]

[23]   Hasanuzzaman M, Fujita M, Oku H, Nahar K, Hawrylak-Nowak B, Eds. Plant nutrients and abiotic stress tolerance. Springer Nature Singapore Pte Ltd. 2018.
[http://dx.doi.org/10.1007/978-981-10-9044-8]

[24]   Khalil U, Ali S, Rizwan M, *et al.* Role of mineral nutrients in plant growth under extreme temperatures In Plant Nutrients and Abiotic Stress Tolerance. Singapore: Springer 2018; pp. 499-524.
[http://dx.doi.org/10.1007/978-981-10-9044-8_21]

[25]   Kang DJ, Endo A, Seo YJ. Effects of soil scraping on the reclamation of tsunami-damaged paddy soil. J Crop Sci Biotechnol 2013; 16(3): 219-23.
[http://dx.doi.org/10.1007/s12892-013-0049-y]

[26]   Yao Z, Li J, Xie H, Yu C. Review on remediation technologies of soil contaminated by heavy metals. Procedia Environ Sci 2012; 16: 722-9.
[http://dx.doi.org/10.1016/j.proenv.2012.10.099]

[27]   Choudhary OP. Use of amendments in ameliorating soil and water sodicity In Bioremediation of salt affected soils: An Indian perspective. Cham: Springer 2017; pp. 195-210.

[28]   Yadav RS, Mahatma MK, Thirumalaisamy PP, *et al.* Arbuscular Mycorrhizal Fungi (AMF) for sustainable soil and plant health in salt-affected soils In Bioremediation of salt affected soils: an Indian perspective. Cham: Springer 2017; pp. 133-56.

[29]   García AC, Izquierdo FG, González OL, *et al.* Biotechnology of humified materials obtained from vermicomposts for sustainable agroecological purposes. Afr J Biotechnol 2013; 12(7). : 625-34.
[http://dx.doi.org/10.5897/AJBX12.014]

[30]   Ondrasek G, Rengel Z, Veres S. Soil salinisation and salt stress in crop production. Abiotic stress in plants-Mechanisms and adaptations 2011; 22: 171-90.
[http://dx.doi.org/10.5772/22248]

[31]   Abou-Baker NH, Abd-Eladl M, Abbas MM. Use of silicate and different cultivation practices in alleviating salt stress effect on bean plants. Aust J Basic Appl Sci 2011; 5(9): 769-81.

[32]   Abd-Eladl M, Fouda S, Abou-Baker N. Bean yield and soil parameters as response to application of biogas residues and ammonium nitrate under different water requirements. Egypt J Soil Sci 2016; 1-14.
[http://dx.doi.org/10.21608/ejss.2016.603]

[33]   Youssef RA, Rasheed MA, Gaber ES, da Silva JA, Abd El Kader AA, Abou-Baker NH. Evaluation of Banana Compost Enriched with Microorganisms on Concentrations of Heavy Metals in Corn and Bean Plants. 2010; 54-8.

[34]   Farooq M, Gogoi N, Hussain M, *et al.* Effects, tolerance mechanisms and management of salt stress in grain legumes. Plant Physiol Biochem 2017; 118: 199-217.
[http://dx.doi.org/10.1016/j.plaphy.2017.06.020] [PMID: 28648997]

[35]   Meena KK, Sorty AM, Bitla UM, *et al.* Abiotic stress responses and microbe-mediated mitigation in plants: the omics strategies. Front Plant Sci 2017; 8: 172.
[http://dx.doi.org/10.3389/fpls.2017.00172] [PMID: 28232845]

[36]   Mishra J, Singh R, Arora NK. Plant growth-promoting microbes: diverse roles in agriculture and environmental sustainability InProbiotics and plant health. Singapore: Springer 2017; pp. 71-111.

[37]   Mehanna HM, Hussein MM, Abou-Baker NH. The relationship between water regimes and maize productivity under drip irrigation system: a statistical model. J Appl Sci Res 2013; 9(6): 3735-41.

[38]   Ibrahim MM, Abd-Eladl M, Abou-Baker NH. Lignocellulosic biomass for the preparation of

cellulose-based hydrogel and its use for optimizing water resources in agriculture. J Appl Polym Sci 2015; 132(42): n/a.
[http://dx.doi.org/10.1002/app.42652]

[39]   Tayel MY, El-Gindy AA, El-Hady MA, Ghany HA. Effect of irrigation systems on: II-yield, water and fertilizer use efficiency of grape. J Appl Sci Res 2007; 3(5): 367-72.

[40]   Abou-Baker NH, Abd-Eladl M, Eid TA. Silicon and water regime responses in bean production under soil saline condition. J Appl Sci Res 2012; 8(12): 5698-707.

[41]   Abd-Eladl M, Abou-Baker NH, El-Ashry S. Impact of compost and mineral fertilization irrigation regime on wheat and sequenced maize plants. Minufiya J Agric Res 2010; 35(6): 2245-62.

[42]   Dewdar MD, Rady MM. Induction of cotton plants to overcome the adverse effects of reclaimed saline soil by calcium paste and ascorbic acid applications. Acad J Agri Res 2013; 1(2): 017-27.
[http://dx.doi.org/10.15413/ajar.2012.0126]

[43]   Mourid SS. Effect of calcium carbonate content on potential toxic heavy metals adsorption in calcareous soils. Curr Sci Int 2014; 3: 141-9.

[44]   Sadeghipour O, Aghaei P. Improving the growth of cowpea (*Vigna unguiculata* L. Walp.) by magnetized water. J Biodivers Environ Sci 2013; 3(1): 37-43.

[45]   Ibrahim A, Mohsen B. Effect of irrigation with magnetically treated water on faba bean growth and composition. Int J Agric Policy Res 2013; 1(2): 24-40.

[46]   Ali TB, Khalil SE, Khalil AM. Magnetic treatments of *Capsicum annuum* L. grown under saline irrigation conditions. J Appl Sci Res 2011; 7(11): 1558-68.

[47]   Zein FI, El-Sanafawy H, Talha NI, Salama SA. Using canola plants for phytoextracting heavy metals from soils irrigated with polluted drainage water for a long term. Journal of Soil Sciences and Agricultural Engineering 2009; 34(6): 7309-23.
[http://dx.doi.org/10.21608/jssae.2009.103832]

[48]   Wu G, Kang H, Zhang X, Shao H, Chu L, Ruan C. A critical review on the bio-removal of hazardous heavy metals from contaminated soils: Issues, progress, eco-environmental concerns and opportunities. J Hazard Mater 2010; 174(1-3): 1-8.
[http://dx.doi.org/10.1016/j.jhazmat.2009.09.113] [PMID: 19864055]

[49]   Mohamed AA, Eichler-Lobermann B, Schnug E. Response of crops to salinity under Egyptian conditions: a review. Landbauforsch Völkenrode 2007; 57(2): 119.

[50]   Tawfik MM, Thalooth AT, Nabila MZ, Hassanein MS, Amany AB, Amal GA. Sustainable production of Kochia indica grown in saline habitat. Enviro Treat Tech 2013; 1(1): 56-61.

[51]   Hasanuzzaman M, Nahar K, Alam M, *et al.* Potential use of halophytes to remediate saline soils. BioMed research international 2014.

[52]   Sharma A, Gontia I, Agarwal PK, Jha B. Accumulation of heavy metals and its biochemical responses in *Salicornia brachiata*, an extreme halophyte. Mar Biol Res 2010; 6(5): 511-8.
[http://dx.doi.org/10.1080/17451000903434064]

[53]   Barakat NAM, Cazzato E, Nedjimi B, Kabiel HF, Laudadio V, Tufarelli V. Ecophysiological and species-specific responses to seasonal variations in halophytic species of the chenopodiaceae in a Mediterranean salt marsh. Afr J Ecol 2014; 52(2): 163-72.
[http://dx.doi.org/10.1111/aje.12100]

[54]   Nedjimi B. Heavy metal tolerance in two Algerian saltbushes: A review on plant responses to cadmium and role of calcium in its mitigation. In Plant Nutrients and Abiotic Stress Tolerance. 2018.
[http://dx.doi.org/10.1007/978-981-10-9044-8_9]

[55]   El-Ramady H, Alshaal T, Elhawat N, *et al.* Plant nutrients and their roles under saline soil conditions.Plant nutrients and abiotic stress tolerance. Singapore: Springer 2018; pp. 297-324.
[http://dx.doi.org/10.1007/978-981-10-9044-8_13]

[56]   Çavuşoğlu K, Kaya F, Kılıç S. Effects of boric acid pretreatment on the seed germination, seedling growth and leaf anatomy of barley under saline conditions. J Food Agric Environ 2013; 11: 376-80.

[57]   Hussein MM, El-Saady AM, Abou-Baker NH. Castor bean plants response to phosphorus sources under irrigation by diluted seawater. Int J Chemtech Res 2015; 8(9): 261-71.

[58]   Secco D, Whelan J, Rouached H, Lister R. Nutrient stress-induced chromatin changes in plants. Curr Opin Plant Biol 2017; 39: 1-7.
[http://dx.doi.org/10.1016/j.pbi.2017.04.001] [PMID: 28441589]

[59]   Khan M, Ahmad R, Khan MD, *et al.* Trace elements in abiotic stress tolerance. In Plant Nutrients and Abiotic Stress Tolerance. Singapore: Springer 2018; pp. 137-51.
[http://dx.doi.org/10.1007/978-981-10-9044-8_5]

[60]   Hussein MM, El-Faham SY, El-Moti EZA, Abou-Baker NH. Jojoba irrigated with diluted seawater as affected by ascorbic acid application. Int J Agric Res 2016; 12(1): 1-9.
[http://dx.doi.org/10.3923/ijar.2017.1.9]

[61]   El-Ramady H, Alshaal T, Abowaly M, *et al.* Nanoremediation for sustainable crop production InNanoscience in food and agriculture 5. Cham: Springer 2017; pp. 335-63.

[62]   Tripathi DK, Ahmad P, Sharma S, Chauhan DK, Dubey NK, Eds. Nanomaterials in Plants, Algae, and Microorganisms: Concepts and Controversies. Academic Press 2017; Vol. 1: pp. 345-91.

[63]   Ojha S, Singh D, Sett A, Chetia H, Kabiraj D, Bora U. Nanotechnology in crop protection.In Nanomaterials in plants, algae, and microorganisms 2018; 345-91. Academic Press

[64]   Ibrahim RK, Hayyan M, AlSaadi MA, Hayyan A, Ibrahim S. Environmental application of nanotechnology: air, soil, and water. Environ Sci Pollut Res Int 2016; 23(14): 13754-88.
[http://dx.doi.org/10.1007/s11356-016-6457-z] [PMID: 27074929]

[65]   Patra AK, Adhikari T, Bhardwaj AK. Enhancing crop productivity in salt-affected environments by stimulating soil biological processes and remediation using nanotechnology InInnovative saline agriculture. New Delhi: Springer 2016; pp. 83-103.

[66]   Kumar V, Guleria P, Kumar V, Yadav SK. Gold nanoparticle exposure induces growth and yield enhancement in Arabidopsis thaliana. Sci Total Environ 2013; 461-462: 462-8.
[http://dx.doi.org/10.1016/j.scitotenv.2013.05.018] [PMID: 23747561]

[67]   Alsaeedi AH, El-Ramady H, Alshaal T, El-Garawani M, Elhawat N, Almohsen M. Engineered silica nanoparticles alleviate the detrimental effects of Na+ stress on germination and growth of common bean (Phaseolus vulgaris). Environ Sci Pollut Res Int 2017; 24(27): 21917-28.
[http://dx.doi.org/10.1007/s11356-017-9847-y] [PMID: 28780690]

[68]   Sabaghnia N, Janmohammadi M. Effect of nano-silicon particles application on salinity tolerance in early growth of some lentil genotypes. In Annales Universitatis Mariae Curie-Sklodowska, sectio C–Biologia 69(2): 39.2015;

[69]   Alloway BJ. Soil factors associated with zinc deficiency in crops and humans. Environ Geochem Health 2009; 31(5): 537-48.
[http://dx.doi.org/10.1007/s10653-009-9255-4] [PMID: 19291414]

[70]   Wang L, Ning C, Pan T, Cai K. Role of Silica Nanoparticles in Abiotic and Biotic Stress Tolerance in Plants: A Review. Int J Mol Sci 2022; 23(4): 1947.
[http://dx.doi.org/10.3390/ijms23041947] [PMID: 35216062]

[71]   Li Z, Zhu L, Zhao F, *et al.* Plant Salinity Stress Response and Nano-Enabled Plant Salt Tolerance. Front Plant Sci 2022; 13: 843994.
[http://dx.doi.org/10.3389/fpls.2022.843994] [PMID: 35392516]

[72]   Kumari S, Khanna RR, Nazir F, *et al.* Bio-Synthesized Nanoparticles in Developing Plant Abiotic Stress Resilience: A New Boon for Sustainable Approach. Int J Mol Sci 2022; 23(8): 4452.
[http://dx.doi.org/10.3390/ijms23084452] [PMID: 35457269]

[73]   Haghighi M, Pessarakli M. Influence of silicon and nano-silicon on salinity tolerance of cherry tomatoes (*Solanum lycopersicum* L.) at early growth stage. Sci Hortic (Amsterdam) 2013; 161: 111-7.
[http://dx.doi.org/10.1016/j.scienta.2013.06.034]

[74]   Haghighi M, Abolghasemi R, Teixeira da Silva JA. Low and high temperature stress affect the growth characteristics of tomato in hydroponic culture with Se and nano-Se amendment. Sci Hortic (Amsterdam) 2014; 178: 231-40.
[http://dx.doi.org/10.1016/j.scienta.2014.09.006]

[75]   Taha RA, Hassan MM, Ibrahim EA, Abou Baker NH, Shaaban EA. Carbon nanotubes impact on date palm *in vitro* cultures. Plant Cell Tissue Organ Cult 2016; 127(2): 525-34. [PCTOC].
[http://dx.doi.org/10.1007/s11240-016-1058-6]

[76]   Hussein MM, Abou-Baker NH. The contribution of nano-zinc to alleviate salinity stress on cotton plants. R Soc Open Sci 2018; 5(8): 171809.
[http://dx.doi.org/10.1098/rsos.171809] [PMID: 30224982]

[77]   Siskani A, Seghatoleslami M, Moosavi G. Effect of deficit irrigation and nano fertilizers on yield and some morphological traits of cotton. 2015.

[78]   Tariq M, Choudhary S, Singh H, *et al.* Role of Nanoparticles in Abiotic Stress. Technology in Agriculture 2021; 323: 1-18.

[79]   Abou-Baker NH, Hussein MM, El-Ashry SM. Comparison between nano iron and iron EDTA as foliar fertilizers under salt stress conditions. Plant Cell Biotechnol Mol Biol 2020; 3: 17-32.

[80]   Stoch L. Crystallochemical aspects of structure controlled processes in oxide glasses. Opt Appl 2008; 38(1).

[81]   Ouis MA, Abd-Eladl M, Abou-Baker NH. Evaluation of agriglass as an environment friendly slow release fertilizer. Silicon 2018; 10(2): 293-9.
[http://dx.doi.org/10.1007/s12633-016-9443-7]

[82]   Abou-Baker NH, Ouis M, Abd-Eladl M. Appraisal of agriglass in promoting maize production under abiotic stress conditions. Silicon 2018; 10(5): 1841-9.
[http://dx.doi.org/10.1007/s12633-017-9684-0]

[83]   Azooz MA, ElBatal HA, ElBadry KM, Abd ElMoneim M, ElAshry SM. Preparation and application of some phosphoborosilicate glasses containing micronutrients as plant fertilisers. Glass Technology-Eur J Glass Sci Tech 2006; 47(6): 164-6.

[84]   Hashesh WM, Habib FM, Wahba MM, Noufal EH, Abou-Baker NH. Removal of lead from polluted water by using clay minerals. Zagazig J Agric Res 2017; 44(6): 2097-103.
[http://dx.doi.org/10.21608/zjar.2017.51245]

[85]   Mubarak DM. Removal of organic and inorganic pollutants from aqueous solutions by organically modified clayey sediments 2012.

[86]   Awad F, Schulz R, Ruse R, Breuer J. Sorption of cadmium and copper by organo-clay complexes. Egypt J Soil Sci 2010; 50(1): 125-39.

[87]   Cruz-Guzmán M, Celis R, Hermosín MC, Koskinen WC, Nater EA, Cornejo J. Heavy metal adsorption by montmorillonites modified with natural organic cations. Soil Sci Soc Am J 2006; 70(1): 215-21.
[http://dx.doi.org/10.2136/sssaj2005.0131]

[88]   Zhang X, Wang X. Adsorption and desorption of nickel(II) ions from aqueous solution by a lignocellulose/montmorillonite nanocomposite. PLoS One 2015; 10(2): e0117077.
[http://dx.doi.org/10.1371/journal.pone.0117077] [PMID: 25647398]

[89]   Abou-Baker NH, El-Ashry MS, Wahba MM, Habib FM, Noufal EHA, Hashesh WM. Evaluating the Application of some Natural Amendments to Hinder the Pollution of Barley Plants by Heavy Metals. Scientific J King Faisal University 2019; 20(2-1441H): 55-68.

[90] Shabeer TPA, Saha A, Gajbhiye VT, Gupta S, Manjaiah KM, Varghese E. Exploitation of nano-bentonite, nano-halloysite and organically modified nano-montmorillonite as an adsorbent and coagulation aid for the removal of multi-pesticides from water: a sorption modelling approach. Water Air Soil Pollut 2015; 226(3): 41.
[http://dx.doi.org/10.1007/s11270-015-2331-8]

[91] Sciascia L, Casella S, Cavallaro G, *et al.* Olive mill wastewaters decontamination based on organo-nano-clay composites. Ceram Int 2019; 45(2): 2751-9.
[http://dx.doi.org/10.1016/j.ceramint.2018.08.155]

[92] Abd El Zaher FH, Lashein A, Abou-Baker NH. Application of zeolite as a rhizobial carrier under saline conditions. Bioscience Research 2018; 15(2): 1319-33.

[93] Abou-Baker NH, Ibrahim EA, Abd-Eladl MM. Biozeolite for improving bean production under abiotic stress conditions. Bull Transilv Univ Brasov For Wood Ind Agric Food Eng Ser II 2017; 10: 31-46.

[94] Abd El-All AA, Elsherif MH, Shehata HS, El-Shahat RM. Efficiency use of nitrogen, biofertilizers and composted biostraw on rice production under saline soil. J Appl Sci Res 2013; 9(3): 1604-11.

[95] Jawor A, Hoek EMV. Removing cadmium ions from water *via* nanoparticle-enhanced ultrafiltration. Environ Sci Technol 2010; 44(7): 2570-6.
[http://dx.doi.org/10.1021/es902310e] [PMID: 20230021]

[96] Asaad JN, Ikladious NE, Awad F, Müller T. Evaluation of some new hyperbranched polyesters as binding agents for heavy metals. Can J Chem Eng 2013; 91(2): 257-63.
[http://dx.doi.org/10.1002/cjce.21634]

[97] Liu P. Nanopolysaccharides for adsorption of heavy metal ions from water 2014.

[98] El-Hady OA, Abou-Sedera SA. The conditioning effect of composts (natural) or/and acrylamide hydrogels (synthesized) on a sandy calcareous soil. II. Chemical and biological properties of the soil. Egypt J Soil Sci 2006; 46: 538-46.

[99] Adil AY, Canan KA, Metin TU. Hydrogel substrate alleviates salt stress with increase antioxidant enzymes activity of bean (*Phaseolus vulgaris* L.) under salinity stress. Afr J Agric Res 2011; 6(3): 715-24.

[100] Koupai JA, Eslamian SS, Kazemi JA. Enhancing the available water content in unsaturated soil zone using hydrogel, to improve plant growth indices. Ecohydrol Hydrobiol 2008; 8(1): 67-75.
[http://dx.doi.org/10.2478/v10104-009-0005-0]

[101] Liang Y, Sun W, Zhu YG, Christie P. Mechanisms of silicon-mediated alleviation of abiotic stresses in higher plants: A review. Environ Pollut 2007; 147(2): 422-8.
[http://dx.doi.org/10.1016/j.envpol.2006.06.008] [PMID: 16996179]

[102] Hamed N, Abdel-Aziz RAH, Abou-Baker N. Effect of some applications on the performance of mandarin trees under soil salinity conditions. Egypt J Hortic 2017; 0(0): 0.
[http://dx.doi.org/10.21608/ejoh.2017.1742.1021]

[103] Hussein MM, Mehanna H, Abou-Baker NH. Growth, photosynthetic pigmentsand mineral status of cotton plants as affected by salicylic acid and salt stress. J Appl Sci Res 2012; (November): 5476-84.

[104] Mroczek-Zdyrska M, Strubińska J, Hanaka A. Selenium improves physiological parameters and alleviates oxidative stress in shoots of lead-exposed Vicia faba L. minor plants grown under phosphorus-deficient conditions. J Plant Growth Regul 2017; 36(1): 186-99.
[http://dx.doi.org/10.1007/s00344-016-9629-7]

[105] Cao F, Fu M, Wang R, Cheng W, Zhang G, Wu F. Genotypic-dependent effects of N fertilizer, glutathione, silicon, zinc, and selenium on proteomic profiles, amino acid contents, and quality of rice genotypes with contrasting grain Cd accumulation. Funct Integr Genomics 2017; 17(4): 387-97.
[http://dx.doi.org/10.1007/s10142-016-0540-x] [PMID: 27999965]

[106] Sattar A, Cheema MA, Abbas T, Sher A, Ijaz M, Hussain M. Separate and combined effects of silicon

and selenium on salt tolerance of wheat plants. Russ J Plant Physiol 2017; 64(3): 341-8.
[http://dx.doi.org/10.1134/S1021443717030141]

[107] Alsokari SS. Synergistic effect of kinetin and spermine on some physiological aspects of seawater stressed Vigna sinensis plants. Saudi J Biol Sci 2011; 18(1): 37-44.
[http://dx.doi.org/10.1016/j.sjbs.2010.07.002] [PMID: 23961102]

[108] Gurmani AR, Bano A, Ullah N, Khan H, Jahangir M, Flowers TJ. Exogenous abscisic acid (ABA) and silicon (Si) promote salinity tolerance by reducing sodium (Na^+) transport and bypass flow in rice ('*Oryza sativa'indica*). Aust J Crop Sci 2013; 7(9): 1219-26.

[109] Carrera J, Rodrigo G, Jaramillo A, Elena SF. Reverse-engineering the Arabidopsis thaliana transcriptional network under changing environmental conditions. Genome Biol 2009; 10(9): R96.
[http://dx.doi.org/10.1186/gb-2009-10-9-r96] [PMID: 19754933]

[110] Ding ZJ, Yan JY, Xu XY, *et al.* Transcription factor WRKY46 regulates osmotic stress responses and stomatal movement independently in Arabidopsis. Plant J 2014; 79(1): 13-27.
[http://dx.doi.org/10.1111/tpj.12538] [PMID: 24773321]

[111] Zhang X, Long Y, Huang J, Xia J. OsNAC45 is involved in ABA response and salt tolerance in rice. Rice (N Y) 2020; 13(1): 79.
[http://dx.doi.org/10.1186/s12284-020-00440-1] [PMID: 33284415]

[112] Qadir M, Quillérou E, Nangia V, *et al.* Economics of salt-induced land degradation and restoration. InNatural resources forum 2014; 38(4): 282-95.

CHAPTER 8

Understanding Molecular Mechanisms of Plant Physiological Responses Under Drought and Salt Stresses

Abhishek Kanojia[1], Ayushi Jaiswal[1] and Yashwanti Mudgil[1,*]

[1] *Department of Botany, University of Delhi, New Delhi-110007, India*

Abstract: The change in global climate patterns raised issues related to soil salinization, desertification, unseasonal rains, and droughts which directly or indirectly influence agricultural produce. Plants have some level of tolerance towards various stresses, and this tolerance capacity varies among plant species based on their genetic constitution and evolutionary adaptability. Abiotic stress sensing and responses in plants involve complex pathways containing multiple steps and genes. To survive in stressful conditions, plants need to adjust their physiological and metabolic processes. Adjustments in these processes involve complex changes at the molecular level resulting in a plant's adaptation at a morphological and developmental level, which in turn impacts agriculture yields (biomass). Here in this chapter, we are emphasizing molecular dissection of the physiological responses towards salt and drought stress. The study of salt and drought stress responses in plants is also important from an agricultural perspective. We aim to provide up-to-date advancements in the molecular biology field to explain 'stress sensing to stress response' in plants which involves multifaceted pathways and networks. We will be covering the process starting from sensing, transfer of signals, regulation of gene expressions, synthesis of osmolytes-metabolites, ROS scavenging pathways, *etc..*, involved in the survival of plants. This chapter will specifically address information regarding salt and drought stress effects and responses in plants.

Keywords: Drought, Environmental Stress, Molecular Physiology, Salt, Signaling.

1. INTRODUCTION

Different plants belonging to various climatic zones have different stress tolerance potentials, which are based on their genetic constitution and evolutionary evolved adaptability. When plants experience stress, they tend to overcome it by changing physiological and metabolic processes which affect growth and development and

---

** **Corresponding author Yashwanti Mudgil:** Department of Botany, University of Delhi, Delhi 110007, India; E-mail: ymudgil@gmail.com*

Jen-Tsung Chen (Ed.)

assist the plants to handle stressful conditions. According to various reports, the agriculture sector would be highly impacted in the near future due to changing climate, which has increased the weather extremities. It has been noted that the number of warms days also increases in a year [1].

Majorly tropical and sub-tropical regions of the world are getting affected by heat stress and interrelated consequences [2]. Salt stress or salinity stress also negatively impacts the overall yield of the crops [3].

There are different reasons for salt stress, like using poor-quality water, which contains a high amount of soluble salts in irrigation [4]. Studies suggest that approximately 800Mha of land is affected by salt stress [5]. It was also reported that the presence of a high amount of salt in soil interferes with the absorption of other nutrients [6]. The major traits affected by salt stress are germination, leaf area, height, reduction in water uptake capacity, root architecture, *etc..* [7 - 10]. In broad terms, salt stress causes osmotic stress and ionic toxicity. If we compare the osmotic pressure in plant cells and soil solution under normal conditions, the plant cells have high osmotic pressure. In salt stress conditions, the osmotic pressure in soil solution becomes more than that of plant cells. This restricts the flow of water into the plants. This flow of important minerals like K^+ and Ca^{2+} is also restricted. To avoid the ill effects of Na^+ and Cl^-, plants generally adopt sequestration of these ions in a cellular compartment like a vacuole.

Drought or water shortage also has a great impact on the agriculture sector. The change in rainfall patterns due to climate change is a major challenge in the agriculture system [11]. Most agriculture depends upon rainfall for irrigation purposes [12]. Common effects are a decrease in plant height, leaf area, leaf curl, small flowers and fruits, leaf senescence, destruction of photosynthetic pigments, and ultimately, effects on the yield of the crops [13 - 16]. The acclimatization of drought stress doesn't work in many crops as many of them are annual and harvested in a single generation only. To cope with drought stress, we need a better variety of seeds that have stress tolerance capacity and remain minimally affected due to stress conditions. Any kind of stress ultimately impacts the yield of crops. The change in the morpho-physiological form of plants can help them to survive the stress, but that creates huge agroeconomic losses.

In this chapter, we will be covering the process starting from sensing and transfer of stress signals, regulation of gene expressions, synthesis of osmolytes-metabolites, ROS (Reactive Oxygen Species) scavenging pathways, *etc.*

1.1. Signaling Mechanisms Under Salt Stress

Till now, the exact mechanism of salt stress perception/sense is not known.

Whether there is any receptor or sensor present on the plasma membrane is also not known. There are different signaling mechanisms present in plants that get activated under stress.

- Calcium-based pathway: It was reported that during salt stress, the cytosolic concentration of calcium ions increases. The rise of calcium ions initiates further downstream transfer of signals [17]. Calcium-dependent protein kinases (CDPKs), calcineurin B-like proteins (CBLs), and CBL-interacting protein kinases (CIPKs) were also reported to regulate gene expression under hyperosmotic stress [18 - 19]. Besides these, various other TFs are known to be directly activated by the calcium ions like MYBs, GTLs (GT element-binding like proteins), Calmodulin-binding transcription activators (CAMTAs), *etc.*. [20 - 22].
- The protein kinases pathway: The activation of Mitogen-Activated Protein Kinases (MAPKs) is well known in the transduction of signals in stress conditions. The activation of the MAPK pathway affects the expression of various transcription factors involved in stress which alter the gene expression. The MAPK cascade is generally activated by ROS generation due to stress [23].
- Cyclic Nucleotides: The level of cGMP under salt stress gets increased and activates two different cascades with and without calcium involvement. It was also shown that cGMP inhibits sodium influx in the cells [24 - 28].
- SOS (Salt overlay sensitive) signaling pathway: SOS-dependent signaling pathway constitutes a major ion homeostasis maintenance system. It is one of the major cell sodium exporters. The discovery of the SOS system allowed a better understanding of salt stress signaling [29].
- ABA-dependent pathway: salt stress also activates the ABA-dependent pathway. It was also noted that the ABA level increases in plant cells when treated with salt and affects downstream signaling [30]. ABA influences the expression of various stress-responsive transcription factors, which in turn regulate the expression of several genes involved in ABA biosynthesis like zeaxanthin epoxidase, z9-cis-epoxy carotenoid dioxygenase, the aldehyde oxidase gene, *etc.* [31 - 32].
- Lipid pathway: Plasma membrane contains many lipid derivatives. Phospholipids, phosphatidic acid, sphingolipids, *etc.* Most lipid-based molecules act as secondary cellular messengers involved in the signaling cascade [33] Fig (**1**).

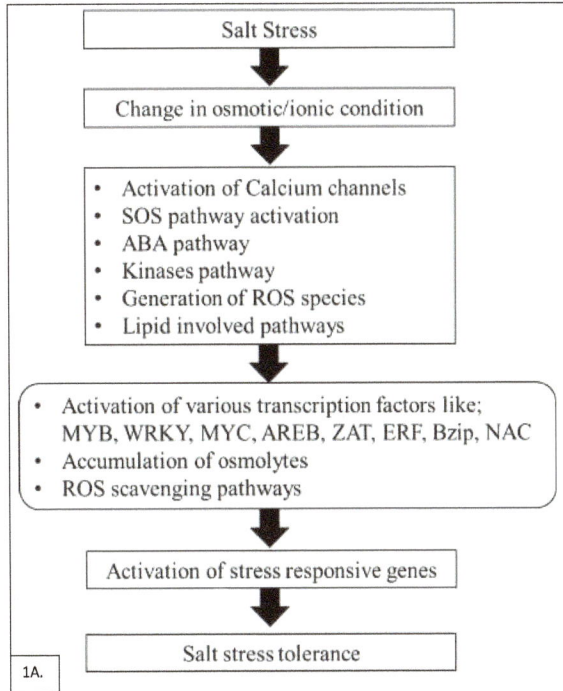

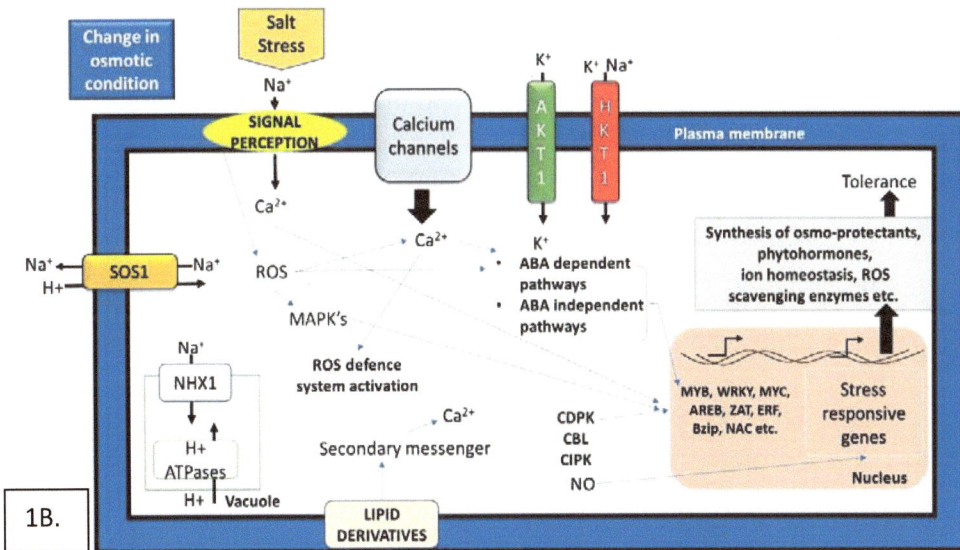

Fig. (1). Simplified known signaling cascade involved in perception and response towards salt stress. (1A.) Flowchart representing signaling cascade in salt stress. (1B.) Known signaling cascade in salt stress in plant cells.

1.2. Salt Stress Regulation in Plants

Plants use different mechanisms which help them to sustain salt stress. The major changes in biochemical and physiological pathways depend on the duration and severity of the stress. The change in expression of genes, *i.e.*, upregulation or downregulation in the influence of salt stress, was also seen in many plants. In *Arabidopsis thaliana*, the transcriptome analysis highlights the induction of approximately 932 genes with the repression of 364 genes [34]. In general, when glycophytic plants experience salt stress, the productivity decreases. The main salt stress-regulated processes are mentioned below:

• Ion transport and compartmentalization
• Osmo-protectants
• Antioxidants pathway
• Hormonal variations
• Nitrogen related molecules

Ion Transport and Compartmentalization: Ion homeostasis plays an important role in the maintenance of different processes in plants. Different plants have varying degrees of salt stress tolerance. SOS sensing and downstream signaling play an important role in ion homeostasis maintenance [35]. The SOS pathway consists of three major proteins SOS1, SOS2, and SOS3. SOS1 is a Na+/H+ antiporter located at the plasma membrane [35 - 36]. SOS2 is a serine/threonine kinase [37]. SOS3 is a myristoylated Ca^{2+} binding protein [38]. SOS1 act as a major salt exporter from the cell [35]. SOS2 and SOS3 have special motifs which aid in their interaction leading to downstream processes. SOS3 and SOS2 proteins interact, which also removes self-inhibition later. The complex of SOS3 and SOS2 phosphorylates SOS1 on the plasma membrane resulting in the efflux of sodium ions and aiding in salt tolerance. Different ion exchange channels present in the plasma membrane play an important role in cellular detoxification [39 - 40]. Sodium/Hydrogen (Na^+/H^+) exchanger (NHX1) present on the tonoplast membrane removes sodium ions from the cytosol and sequesters them in the vacuole [41 - 42]. *NHX1* overexpression-based studies also revealed that it increases salinity tolerance in plants like *Arabidopsis*, tomato, and rice [43 - 44]. Potassium channel AKT1 also plays an important role in maintaining the K^+/Na^+ ratio in plant cells [45].

Osmo-protectants: Plants have different osmolytes, which help in the smooth functioning of cellular metabolism. The major compounds are proline, betaine, polyols, *etc.*. These molecules are deprived of any charges and are polar. Proline is one of the widely studied osmolytes; it's concentration increases under different stresses, including salt stress [46 - 47]. It occurs in varied plant groups. Glutamate or ornithine acts as a precursor of proline. Major genes involved in its synthesis are pyrroline-5-carboxylate synthetase (P5CS), pyrroline-5-carboxylate (P5C), and P5C reductase (P5CR) [48]. The study showed that proline also acts as a singlet oxygen quencher and ROS scavenger, highlighting its antioxidant property [49]. They also perform the function of molecular chaperons by maintaining protein integrity [50]. A study in olives shows that the external application of proline helps plants in tolerating salt stress by enhancing antioxidant pathways and improving photosynthetic machinery [51]. A study in *Nicotiana tabacum* shows that the application of proline enhances the activity of enzymes involved in the antioxidant defense system [52]. In Rice also, exogenous application of low-quantity proline ameliorates salt stress [53]. In *Arachis hypogea*, the low concentration of proline enhances the fresh weight, which declines due to salt stress [54]. Betaine, better known as glycine betaine, is another compound that acts as an osmoprotectant [55]. It's a quaternary ammonium compound that is amphoteric in nature and ubiquitously present in plants. It also aids in protein stabilization, photosynthetic system protection, and ROS reduction [56]. Choline or glycine acts as a precursor for the production of glycine betaine. In plants, more than one pathway of betaine generation is present [57]. Study shows that betaine application to rice plants mitigates damages induced due to salt-stressed [58]. The foliar spray of betaine also enhances the photosynthetic rate due to pigment stabilization under stress conditions [59]. Sugar alcohols like polyols are compounds with multiple hydroxyl groups acting as osmoprotectants [60]. Mannitol, pinitol, sorbitol, and inositol are important examples of polyols. Carbohydrate sugars also accumulate in plants when exposed to salt stress [61]. It is also reported that the concentration of sugars like fructose and sucrose increased in plant cells while this trend fluctuates and varies from species to species [62].

Hormonal Variation: ABA accumulation in stress conditions is widely known. The accrual of ABA helps in the mitigation of salt stress on growth and assimilates translocation [31, 63]. Study shows that an increase in ABA concentration also aids in the increase of solutes like proline, sugars, and ions like Ca^{2+} and K^+ in the vacuoles of roots which leads to salt stress tolerance [64]. ABA also acts as an important signaling molecule that leads to the activation of various stress-responsive genes involved in salt stress. Meanwhile, the expression of genes involved in ABA synthesis also increases under salt stress, like 9-cisepoxycarotenoid dioxygenase, ABA-aldehyde oxidase, zeaxanthin oxidase, *etc.*., which may occur by calcium-dependent signaling pathways [65]. Studies in the rice system indicate that the expression of some genes like *OsDSM2* (Drought-hypersensitive mutant2) and *OsCam1*-1 (Ca2+-binding calmodulin) leads to a higher accumulation of ABA as these genes were found to be involved in ABA biosynthesis and calcium-binding respectively [66]. In rice, the high ABA level in salt stress also leads to the upregulation of MAPKs. Some studies indicate that the expression of salt-responsive genes can be dependent or independent of ABA [67]. It was reported that salt treatment leads to changes in root architecture, like inhibition of lateral root formation by varying the auxin distribution and accumulation [68 - 69]. The overexpression of the *YUCCA3* gene leads to hypersensitivity towards salt stress [70]. A study on tomatoes also showed a reduction in IAA concentration under salt stress [71]. Jasmonate (JA) accumulation was also reported in plants under salt stress [72]. Furthermore, in some crops, external application of JA leads to the recovery of seedlings affected the salt stress and it also leads to reduced Na^+ in salt-tolerant rice [73 - 74]. This indicates that a high level of JA may be helpful in salt stress tolerance. Gibberellins (GA3) application increases seed germination, and grain yield under salt stress [75]. It indicates that GA might be involved in reducing the effects of salt stress. Studies also indicate ethylene's role in response to salt stress. Mutant analysis of ethylene receptor *etr1*-7 showed better tolerance towards salt stress in *Arabidopsis* [76 - 77]. Also, overexpression of ethylene insensitive 3 EIN3, the central regulator in ethylene signaling shows better tolerance to salt stress in terms of germination and growth [78]. While the *ein3-1* and *ein2-1* mutants show hypersensitivity towards salt stress [76]. The overexpression of ethylene receptor (NTHK1) of *Nicotiana tabacum* in *Arabidopsis thaliana,* showed upregulation of salt stress-responsive genes like COR6.6, RD17, ERF4, RD21A, and VSP2 [79 - 80]. Collectively these studies indicate that ethylene-related pathways may be involved in salt stress response.

Antioxidant Pathways: Abiotic stress, especially salt stress, disrupts the electron transport chains (*ETC.*) and other cellular processes, which leads to the generation of ROS. Cell chloroplast act as a major source of ROS generation; under salt stress, ROS generation increases due to changes in membrane fluidity and the formation of protein complexes along with blockage of electron transfer from water to PSII [81 - 84]. Mitochondrial respiration also gets disturbed under salt stress which leads to electron leakage due to the reduction of ubiquinone pool of complex I and III in *ETC.* to oxygen which results in the generation of superoxide anion [85 - 86]. Besides, surplus O_2 upsurges the photorespiration rate, which produces other ROS as a by-product [87 - 88].

There are different mechanisms and sources of ROS generation. ROS molecules have strong oxidizing potential, which is detrimental to cell integrity. Antioxidant metabolism pathways present in plants can involve enzymatic or non-enzymatic paths. These molecules are known to detoxify different ROS forms. In the non-enzymatic antioxidant system, compounds like tocopherols, carotenoids, and phenolic compounds are involved. Tocopherols are lipophilic antioxidants and prevent lipid oxidation. Analysis of different plant species showed an increase in salt stress tolerance due to an increase in tocopherol concentration. The overexpression of rice tocopherol cyclase gene (OsVTE1) in rice showed better tolerance to salt stress and produced a lesser amount of hydrogen peroxide [89 - 90]. Carotenoids are also classified as lipophilic antioxidants, research on sugarcane varieties shows that salt-tolerant varieties have more amount of carotenoid content compared to other varieties [91]. Tannins, lignin, flavonoids, hydroxycinnamate esters, *etc..*, are phenolic compounds having antioxidant properties [92]. Flavonoids and phenylpropanoids are involved in the scavenging of ROS in combination with peroxidase and phenolic/ascorbic acid pathways. It was also reported that potatoes enriched with flavonoids showed enhanced antioxidant capacity [93]. The enzymatic pathway consists of different enzymes like SOD, CAT, GSH, and ascorbate. The overproduction of enzyme-like SOD leads to better tolerance to salinity [94]. The activity of these enzymes was also reported to increase wheat tolerance to salinity stress comparatively [95]. The expression of genes Mn-SOD, APX, GR, and MDHAR increased in salt-tolerant pea varieties in comparison to a salt-sensitive variety [96]. Transgenic varieties of plants overexpressing antioxidant pathways related to enzymatic genes showed better tolerance towards salt stress and less deterioration due to ROS [97].

Nitrogen-Related Molecules: Nitric oxide (NO) is a gaseous molecule involved in various growth and developmental processes [98]. It was reported that NO is involved in the activation of different genes of the antioxidant pathway [99]. NO molecule acts as a signaling molecule in salt stress and induces salt stress tolerance by enhancing PM H^+-ATPase activity which also involves other signaling molecules [100,101].

The native level of polyamines was reported to get high during abiotic stress conditions [103]. There are different polyamines in plants like spermidine (Spd, N -3-aminopropyl-1, 4-diaminobutane), spermine (Spm, bis (N -3-aminopropy-)-1,4-diaminobutane), putrescine (Put, 1, 4- diaminobutane), cadaverine (Cad, 1, 5-diaminopentane) [105, 106]. The synthesis of different polyamines seems to be interconnected. Ornithine acts as a substrate for putrescine synthesis by the action of ornithine decarboxylase. Besides this, one more path is known for the putrescine synthesis by the sequential action of enzymes arginine decarboxylase (ADC), agmatine iminohydrolase (AIH), and N-carbamoyl-Put amidohydrolase (CPA) [107]. Putrescine also acts as a substrate for spermidine synthesis by the action of spermidine synthase. Spermidine, on the other hand, acts as a precursor for spermine on the action of spermine synthase [108]. Polyamines have a unique structure, polycationic at physiological pH, and are postulated to be involved in the scavenging of free radicals [102]. Polyamines show dual effects on plants during stress.

1.3. Signaling in Drought Stress

The decrease in the water content of the soil changes the water status of leaves and affects plants' physiology [109]. The decrease in turgor pressure and water potential of the leaf affects the synthesis, distribution, and transport of hormones. The change in turgor pressure caused by the loss of water from the cell is perceived as water stress by the cell. This type of signal is called a hydraulic signal [110].

Fromm and co-workers proposed that in drought-stressed plants, communication between shoots and roots is maintained *via* electrical signals [111]. When plants experience the initial water-deficit stress signal, the membrane receptor converts the signal into a chemical signal, triggering the production of the second messenger, mainly Ca^{2+}, ROS, plant hormones, IP_3, and phosphatidic acid signals [110].

Plant hormones are chemical signals that regulate plants' growth and development when present in low concentrations. When the water content of soil decreases, the content of some hormones increases in the plant, which is known as a positive signal. ABA, IAA, and ethylene are examples of positive signals as their content increases under drought stress, whereas cytokinin content decreases, which is an example of a negative signal [109].

ABA is an important signal molecule during drought stress. In addition to promoting aging, inhibiting seed germination, and regulating plant growth, it regulates the response of plants to external stresses, marked by an increased content in the plant during drought, temperature, salinity, and other extreme conditions. ABA forms information connections between the underground and aboveground parts of the plant. During drought stress in plants, the root cells produce ABA that transmits a positive signal to aboveground tissues and organs of the plant to reduce the loss of water through stomatal closure. ABA can induce a wide variety of downstream effectors such as phosphatases, kinases, proteins involved in the ubiquitin pathway, and G-proteins in response to drought stress. ABA has many receptors (PYR/PYL/RCARs) having protein kinase activity which, upon binding to ABA molecules, get activated by changing the structure of proteins, thus activating/inhibiting the downstream signals for transmitting signals between the cells [109].

Being an essential element in plants, calcium maintains the structure of the cell wall, cell membrane stability, intracellular homeostasis, and plant growth and regulates plant development. Extracellular Ca^{2+} activates the increase of intracellular Ca^{2+} concentration *via* a calcium-sensing receptor (CAS), present in the guard cells of *A. thaliana*, confirming their role as the first messenger [112].

Plants synthesize ABA in response to drought to minimize water loss *via* stomatal closure. Cyclic adenosine -5- diphosphate ribose (cADPR), Ca^{2+} and IP3 act as second messengers. ABA increases the concentration of cADPR and IP3, which in turn release Ca^{2+} in guard cells, increasing their concentration in guard cells. CDPKs, CaM, and CBLs recognize Ca^{2+} and adapt to drought stress *via* downstream signal transduction.

ROS acts as a signal that activates the defense system and reduces abiotic stress-induced damage [114]. Drought stress results in stomatal closure, which eventually leads to limited CO_2 fixation and increased ROS production. This enhanced ROS acts as an alarm for the plant that activates the defense pathway in response to drought stress. Plant organelles respond differently to REDOX signals during drought stress Among all ROS, H_2O_2 is one of the most stable ROS that regulates the mobilization of Ca^{2+}, expression of genes, and protein phosphorylation. Yan and co-workers (2007) found that ABA promotes ROS production that acts as a signal in regulating stomatal closure [115]. H_2O_2 induces the phosphorylation of MAPK, regulates gene expression, and is involved in multiple signaling cascades [116].

1.4. Pathways in Details

Plants have evolved highly complex and diverse signaling pathways in drought stress response as it involves various stimuli like oxidative burst, osmotic shock, *etc.*, along with changes in signaling molecules [117] Fig (**2**).

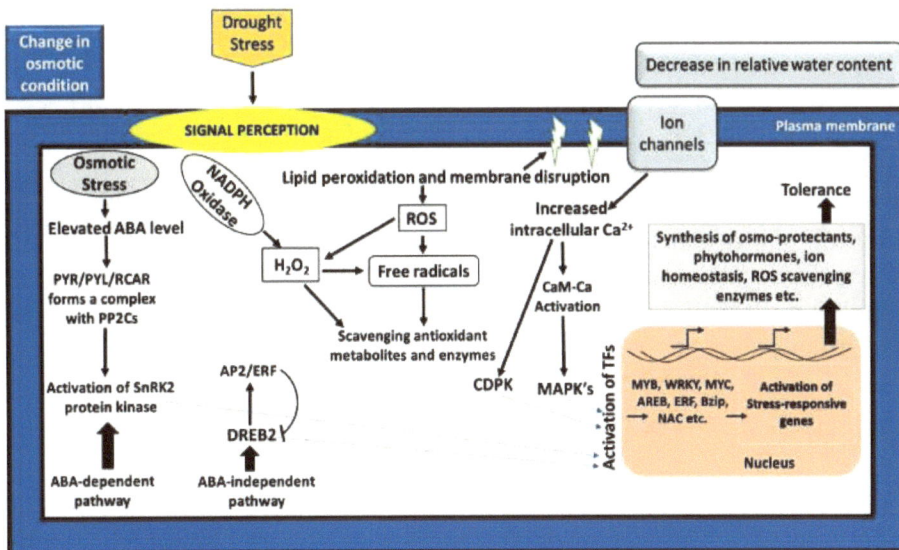

Fig. (2). Simplified known signaling pathway in drought stress.

1.5. The Core ABA-Signalling Pathway

ABA is a key player in drought stress response in plants as it plays a significant role in stomatal closure and gene expression in response to stress. ABA-dependent, as well as ABA-independent pathways, are well-characterized pathways involved in drought stress response as they regulate transcription and gene expression.

In ABA-dependent pathways, TFs bind to ABA-responsive elements (ABRE), while in ABA-independent pathways, TFs bind to dehydration-responsive element/C-repeat (DRE/CRT) in response to drought stress. 2C-type protein phosphatase (PP2C), the PYRABACTIN RESISTANCE / PYR1-LIKE/ REGULATORY COMPONENTS OF ABA RECEPTORS (PYR/PYL/RCAR) and SNF1-related protein kinase 2 (SnRK2) are the three major components of early ABA signaling. ABA as a ligand is recognized by the hydrophobic pocket in the proteins of the ABA receptor, which upon binding, changes the conformation of the receptor, enabling their interaction with group A PP2Cs and inhibiting their phosphatase activity [118].

1.6. PP2C: Regulator of ABA Signalling in Plants

ABI1 and ABI2 are ABA-insensitive (ABI) *Arabidopsis* mutants that encode a homologous PP2C. Protein phosphorylation is important for ABA signaling. ABA hypersensitivity results from the disruption of ABI genes. Thus, group A PP2Cs are negative regulators or repressors of ABA signaling [119].

1.7. ABA Receptors

In 2009 using two different approaches, the mechanism of how group A PP2Cs were involved in ABA signaling was clarified by the identification of soluble ABA receptors, PYR/PYL/RCARs. PYR1 was identified using an ABA agonist by Cutler's group in 2009 by chemical genetics. RCAR1 was identified by Grill's group by the identification of proteins interacting with PP2C. PYR/PYL/RCAR family has 14 members in *Arabidopsis*. ABA as a ligand is recognized by the hydrophobic pocket in each PYR/PYL/RCAR protein. Binding to ABA changes the conformation of PYR/PYL/RCAR protein, enabling their interaction with group A PP2Cs and inhibiting their phosphatase activity [120].

1.8. SnRK2

The SnRK2 family is a protein kinase activated by ABA, comprising ten members in *Arabidopsis*. They are categorized into three subclasses, of which subclass III SnRK2s, *i.e.*, SRK2E/OST1/SnRK2.6, SRK2D/SnRK2.2, and SRK2I/SnRK2.3, are ABA-activated and plays a significant role in response to ABA as their triple-knockout mutant (srk2dei or snrk2.2/2.3/2.6) exhibited extreme ABA insensitivity and impaired most ABA responses, concluding that the subclass III of SnRK2s is positive regulators for ABA signaling [119].

1.9. ABA-Dependent Signalling Pathway

The mechanism of ABA-dependent SnRK2 activation was determined by a series of biochemical studies. Studies revealed that ABA-activated subclass III SnRK2s, interact physically with group A PP2Cs. Group A PP2Cs repress subclass III SnRK2s by dephosphorylating and inactivating subclass III SnRK2s [119].

Five classes of TFs control ABA-dependent gene induction during drought stress. The basic Leucine Zipper Domain (bZIP-type) TFs bind to the ABA response element (ABRE), having ACGTGG/TC consensus. During drought stress and ABA treatment, AREB1/ABF2, AREB2/ABF4, and ABF3 bZIP TFs are expressed in *Arabidopsis*. ABA accumulation and the induction of an ABA-responsive protein kinase are required for the activation of TF by phosphorylation. Drought stress and ABA treatment induce three genes that encode the NAC domain proteins ANAC019, ANAC055, and ANAC072. Transcriptional repressors (Cys2/His2-type zinc-finger proteins), MYB, MYC, and homeodomain TFs also play an important role during ABA response to drought stress. Responsive to Dehydration 22 (RD22) well studied in *Arabidopsis* is ABA inducible gene that contains MYC (CANNTG) and MYB (C/TAACNA/G) recognition sites in its promoter region. MYC and MYB accumulation is observed only after the ABA concentration increases, and their over-expression results in increased ABA sensitivity and drought tolerance [121].

1.10. ABA-Independent Pathway

Developmental cues or environmental stimuli lead to responses such as an elevation in the endogenous levels of ABA, which upon PYR/PYL/RCARs receptor binding interacts with group A PP2Cs and inactivates them which leads to activation of subclass III SnRK2s and phosphorylates TFs such as AREB/ABFs and ion channels, inducing physiological and cellular ABA responses [119]. In the absence of ABA, subclass III SnRK2s are inactivated due to direct dephosphorylation by group A PP2Cs [119].

This pathway is controlled by dehydration response element-binding protein (DREB), containing a DNA binding motif identified originally in APETALA2 (AP2) which is a flower patterning protein. In plants, DREB induces genes related to abiotic stress and confers resistance to them. DREB TFs are categorized into two groups- DREB1 and DREB2. DREB1 plays important role in signaling pathways during low-temperature stress whereas, DREB2 is involved during drought stress. A conserved ERF/AP2 DNA-binding domain is present in DREB proteins [122]. TFs that contain ERF/AP2 DNA-binding domain were identified in various plants like *Arabidopsis* [123], tomato [124], tobacco [125], rice [126,127], and maize [128].

It was found that during drought stress, Early Responsive to Dehydration Stress 1 (ERD1) gene transcripts accumulated before the increase in ABA content, leading to the identification of another ABA-independent pathway [129]. The increased tolerance to drought may be due to the reduced rate of transpiration, *i.e.*, increased stomatal closure. During drought stress, genes such as aquaporins, ERD10, ERD13, and ERF are down-regulated [130]. In *Arabidopsis,* consistent expression of AtMYB61 in the guard cells regulates the functioning of the stomatal aperture [131]. In response to drought, the expression of the Stress Responsive NAC1 (SNAC1) gene is strongly induced inside guard cells, suggesting its role in the closure of stomata [132].

1.11. Early Osmotic Stress Signalling Pathway

Osmo-sensors and osmotic-stress signaling was first identified in the 1990s in yeast cells, in which, Sln1 regulates the Hog1 pathway, a MAPK cascade, that directly controls gene expression in response to osmotic stress. AHK1/AtHK1 were identified as osmo-sensors in *Arabidopsis* because they complement Sln1 in the yeast cells. In response to osmotic stress, protein kinases such as SnRK2s and all its subclasses are also activated. SRK2C/SnRK2.8, belonging to subclass II SnRK2, are positive regulators of drought tolerance in *Arabidopsis*. Plants that overexpress SRK2C, showed increased expression of DREB1A/CBF3 TFs in response to drought stress. A recent study establishing the role of SnRK2s during osmotic-stress signaling in *Arabidopsis* showed that the decuple mutants developed for all ten SnRK2s were hypersensitive to stress suggesting that SnRK2s positive regulate the tolerance of plants to osmotic stress [133]. Both ABA, as well as osmotic stress, can activate SnRK2s but only the ABA-dependent activation mechanism of SnRK2 is established. Osmotic-stress signaling is more rapid than the ABA-dependent signaling pathway as it involves numerous SnRK2 activation mechanisms [133].

1.12. Calcium Dependent Signalling

Ca^{2+} is an important second messenger that affects different protein kinases, including the CDPK family and it plays an important role in both ABA-dependent/independent pathways. Ca^{2+} binding regulates CDPK activity. Osmotic or ABA stress causes an increase in intracellular Ca^{2+}, which affects CDPK activity. CDPKs play an important role during osmotic stress as well as ABA signaling in plants. They regulate stomatal closure and control gene expression. CDPKs are possibly involved in the core ABA pathway in addition to SnRK2 as they share substrates during ABA signaling. AREB/ABFs are important SnRK2 substrates in ABA signaling that can be phosphorylated by AtCPK4 and AtCPK11, *Arabidopsis* CDPKs that positively regulate ABA responses. SLAC1 is another substrate of SnRK2 which plays an important role in stomatal closure as it regulates ion concentration in the guard cells. Studies have reported that AtCPK6, AtCPK21, and AtCPK23 are involved in the phosphorylation of SLAC1. CIPK/SnRK3 is a Ca^{2+} regulated family of protein kinase which plays an important role in ABA as well as osmotic-stress signaling. Ca^{2+} regulates the activity of CIPK and its activity is inhibited in the absence of Ca^{2+}, by autoregulation. Elevation in intracellular Ca^{2+} results in the binding of CBL proteins to Ca^{2+}, followed by their association with CIPKs to mediate their autoregulation [133].

1.13. MAPK-mediated Signalling Pathway

The MAPK activity is regulated by protein kinases such as MAPK, MAPKK, and MAPKKK. This MAPK cascade is conserved in all eukaryotes. MAPK kinases activate MAPKs by phosphorylating the TxY motif, whereas the MAPK phosphatases dephosphorylate the TxY motif and inactivate MAPKs. Plant MAPKs are involved in various signaling pathways, such as in responses to hormones, pathogens, and cell differentiation. ABA or osmotic stress activates several MAPK cascades, such as AtMPK1, 2, 3, and 6. There is limited information about the molecular mechanism of MAPK signaling in response to ABA or osmotic stress, but there are several ways by which MAPKs can be activated by ABA or osmotic stress. ROS activates plant MAPKs and calmodulin(s) can directly regulate MAPK(s). SnRK2 activates MAPK1 and MAPK2 during osmotic stress response [133].

1.14. Proteolysis

It is a regulatory mechanism in signal transduction as it controls the number of signaling factors at the protein level. Target protein ubiquitination of the target proteins by the E3 ubiquitin ligases is responsible for such selective proteolysis. Several E3 ligases involved in ABA signaling are RING E3 ligases, like RHA2a/b, AIP2, SDIR1, and KEEP ON GOING (KEG), *etc.* exist in plants. SUMOylation is also involved in ABA as well as osmotic stress signaling. AtSIZ1, a SUMO ligase, has multiple functions in ABA signaling, phosphate deficiency, and flowering. To have a further understanding of these processes, target proteins are to be identified for both ubiquitination and SUMOylation [133].

1.15. Phospholipid Signalling

Phospholipid signaling is a second messenger involved during ABA as well as osmotic-stress signaling. PA, inositol polyphosphates, and diacylglycerol are phospholipid-derived signal messengers that are generated *via* enzymes like phospholipases, phosphatases, and lipid kinases by membrane lipid modification. Phospholipase D (PLD) hydrolyses phospholipids to generate PA, having an important role in response to drought stress. PLDs and PAs have diverse functions. AtPLDα1 plays an important role in response to ABA and AtPLDδ is involved in response to ROS. PA can bind to several targets like NADPH oxidases, ABI1 (PP2C), and phosphoinositide-dependent protein kinase 1 (PDK). PA can activate MAPKs and/or H$^+$-ATPase The functional role of PAs or PLDs in stress signaling is yet to be discovered [133].

1.16. ROS-mediated Signalling

ROS functions as a signal mediator during ABA as well as osmotic stress signaling. Inside the plant cells, ROS is generated by NADPH oxidase. In *Arabidopsis*, AtrbohD and AtrbohF, are two NADPH oxidases involved in response to ABA in guard cells. ROS attacks proteins, affects signal transduction, and induces functional changes. So, identifying the target proteins for ROS is important. Biochemical studies revealed that group A PP2Cs, ABI1, and ABI2, are H$_2$O$_2$ sensitive at the protein level. NO was discovered in an animal blood vessel. It functions as an endogenous signaling molecule in plant cells. In the guard cells, NO generated by nitrate reductase affects ion channels and is therefore required for stomatal closure. NO stimulates activation of MAPK and ROS activation, suggesting that NO is involved in various abiotic stress, response to phytohormones, and also in response to developmental cues. It is important to identify NO targets in plant cells to enhance our knowledge of Nitric oxide-mediated signaling in plants [133].

1.17. Ethylene (ET) Signalling

Ethylene biosynthesis involves three steps. Methionine is converted to S-adenosyl methionine (SAM). ACC synthase (ACS), along with ACC oxidase (ACO), converts SAM to ethylene. Drought stress negatively regulates ACS, resulting in the reduction of stress-induced ethylene levels during drought stress [134]. In the ethylene signal transduction pathway, drought stress induces ethylene response factors (ERFs), which play an important role in abiotic stress tolerance. ERFs are a member of the APETALA2/ERF family TF. There is an increased expression of ERFs under drought stress and ERFs bind DREs to regulate abiotic stress tolerance. ERFs induce the expression of responsive to desiccation 29B (RD29B), P5CS, germin-like protein 9, early response to dehydration (ERD7) and RD20, osmotin 34, similar to RCD one 5 (SRO5) to provide drought tolerance [135].

1.18. Jasmonic Acid (JA) Signalling

Oxidation of linolenic acid by lipoxygenases *via* the octadecanoid pathway leads to the synthesis of JA [136]. Jasmonyl-isoleucine (JA-Ile) and jasmonyl-isoleucine (JA-Ile) are conjugates of JA [137]. Jasmonic acid is essential for developmental processes, synthesis of secondary metabolites, and the defense responses of plants against various pathogens and wounding [138].

Jasmonic acid enhances drought tolerance by minimizing water loss *via* transpiration and regulating stomatal closure. During drought stress, MeJA increases water uptake and hydraulic conductivity of roots [139]. During drought stress, JAs enhance the activity and expression of drought-induced antioxidant enzymes like catalases, peroxidases, and superoxide dismutase thus playing an important in signaling [140]. Jasmonate ZIM domain (JAZ) proteins are repressors of JA signaling. Under drought stress, JAZ proteins interact with and repress transcription of the target of eat 1 (TOE1) and TOE2 TFs of the Apetala 2 group [141]. 12-OPDA also plays a significant role in JA signaling and drought stress response [142].

1.19. Salicylic Acid (SA) Signalling

Salicylic Acid also plays a significant role in plant development and growth, in response to pathogenesis, biotic and abiotic stresses [143]. Plants resistant to drought stress have increased production of SA levels, while plants that are susceptible show reduced SA levels. In plants, SA synthesis occurs in plastids *via* the phenylalanine ammonia-lyase and isochorismate pathway, both beginning with chorismic acid [144]. The conversion of chorismic acid to isochorismate is carried out by isochorismate synthase [145].

Plants protect themselves from oxidative stress induced by drought, by producing proteins like ascorbate peroxidase (APX), 2-cysteine peroxiredoxin, glutathione S-transferases, and accumulate increased levels of SA [146]. During drought stress, either exogenous application or increased endogenous SA promotes stomatal closure to minimize water loss, resulting in decreased photosynthesis in response to SA-induced ROS accumulation [147].

1.20. Brassinosteroids (BRs) Signalling

Brassinosteroids are plant hormones involved in response to stress in addition to their involvement in plants' growth, productivity, and reproduction [148]. The enhanced tolerance of plants to drought stress, overexpressing BR biosynthetic gene that encodes cytochrome p450, demonstrated that BRs participate in drought stress response [149]. BRASSINOSTEROID INSENSITIVE 1 (BRI1) are receptor proteins that recognize BRs. BRI interacts with BAK1 (BRI-associated receptor kinase 1) co-receptor and regulates gene expression by various phosphorylation and dephosphorylation events [150] Table **1**.

Table 1. Effects of salt and drought stress on common important plants.

Some Common Effects of Salt Stress			
S.No.	**Crop Plants**	**Effects**	**References**
1.	*Vigna unguiculata*	Reduced dry mass, lateral branches	[151]
2.	*Solanum lycopersicum cv. Microtom*	Reduced chlorophyll content, increase in phenol	[152]
3.	*Cicer arietinum*	Decrease in dry mass	[153]
4.	*Fragaria × ananassa*	Increase in root-to-shoot ratio	[154]
5.	*Aloe Vera*	Reduction in leaf length, leaf weight, gel weight, root length	[155]
6.	*Triticum aestivum*	Reduced seedling respiration	[156]
7.	*Gossypium hirsutum*	reduced plant height, fresh and dry weights	[157]
8.	*Zea mays*	reduced plant growth	[158]
9.	*Sorghum bicolor*	Decrease in nutritive quality	[159]
10.	*Brassica rapa*	Retarded root and shoot growth	[160]
11.	*Helianthus annuus*	Reduced growth	[161]
12.	*Oryza sativa*	Reduced germination	[162]
Some common effects of drought			
1.	*Triticum spp.*	Reduced leaf area, relative water content, dry weight, reduced rate of transpiration, photosynthesis, and stomatal conductance.	[163]
2.	*Coffea canephora*	Reduced rate of transpiration, stomatal conductance, decrease in carbon assimilation, and increase in cell wall rigidity.	[164]
3.	*Jatropha curcas*	Increased activity of catalase and accumulation of proline, amino acids, and soluble sugars.	[165]
4.	Field bean, maize	Synthesis of phenolic compounds.	[166]
5.	*Hordeum spontaneum*	Oxidative damage.	[167]
6.	*Gossypium hirsutum*	Increased MDA content and osmolytes, increased root/shoot ratio.	[168]
7.	*Brassica carinata*	Change in root and shoot development, decrease in nitrogen use efficiency.	[169]

(Table 1) cont.....

		Some Common Effects of Salt Stress	
8.	*Beta vulgaris*	Accumulation of glycine betaine, fibrous root production.	[170]
9.	*Jatropha curcas*	Restricted leaf growth, chlorophyll content, stomatal conductance.	[171]
10.	*Populus przewalski*	Reduced growth, productivity, and photosynthetic efficiency.	[172]
11.	*Sophora davidii*	Decrease in the content of photosynthetic pigments and leaf area.	[173]

CONCLUSION

Climate change adversely affects the growth, development, and productivity of plants as they face various kinds of biotic and/or abiotic stresses. Environmental factors responsible for these changes include salinity, drought, change in temperatures, rising CO_2 levels, *etc.* Drought and salinity stress are two prevalent abiotic stresses to which crops are routinely exposed, hindering plant growth, survival, yield, and productivity. In response to abiotic stress, plants activate various stress tolerance mechanisms to adapt to the existing stress conditions. Plants have adapted to these stresses by activating various physiological, morphological, and molecular mechanisms. It is important to understand the mechanism of signal perception by plants, and transmit these perceived signals to their cellular machinery for the activation of transcription factors and gene expression in response to stress to confer stress tolerance. In the current review, we have tried to generate a detailed understanding of the mechanism of drought and salinity stress signaling at the molecular level.

REFERENCES

[1]   Zaval L, Keenan EA, Johnson EJ, Weber EU. How warm days increase belief in global warming. Nat Clim Chang 2014; 4(2): 143-7.
[http://dx.doi.org/10.1038/nclimate2093]

[2]   Teixeira EI, Fischer G, van Velthuizen H, Walter C, Ewert F. Global hot-spots of heat stress on agricultural crops due to climate change. Agric For Meteorol 2013; 170: 206-15.
[http://dx.doi.org/10.1016/j.agrformet.2011.09.002]

[3]   Yadav SP, Bharadwaj R, Nayak H, Mahto R, Singh RK, Prasad SK. Impact of salt stress on growth, productivity and physicochemical properties of plants: A Review. Int J Chem Stud 2019; 7: 1793-8.

[4]   Rengasamy P. Soil processes affecting crop production in salt-affected soils. Funct Plant Biol 2010; 37(7): 613-20.
[http://dx.doi.org/10.1071/FP09249]

[5]   Munns R, Tester M. Mechanisms of salinity tolerance. Ann Rev Plant Biol 2008; 590(651).

[6]   Nawaz K, Muhammad A. Improvement in salt tolerance of maize by exogenous application of glycinebetaine: growth and water relations. Pak J Bot 2007; 39: 1647-53.

[7]   Foolad MR, Lin GY. Genetic potential for salt tolerance during germination in Lycopersicon species. HortScience 1997; 32(2): 296-300.
[http://dx.doi.org/10.21273/HORTSCI.32.2.296]

[8]   Munns R. Comparative physiology of salt and water stress. Plant Cell Environ 2002; 25(2): 239-50.
[http://dx.doi.org/10.1046/j.0016-8025.2001.00808.x] [PMID: 11841667]

[9]   Maggio A, Raimondi G, Martino A, De Pascale S. Salt stress response in tomato beyond the salinity tolerance threshold. Environ Exp Bot 2007; 59(3): 276-82.
[http://dx.doi.org/10.1016/j.envexpbot.2006.02.002]

[10]   Hniličková H, Hnilička F, Orsák M, Hejnák V. Effect of salt stress on growth, electrolyte leakage, Na^+ and K^+ content in selected plant species. Plant Soil Environ 2019; 65(2): 90-6.
[http://dx.doi.org/10.17221/620/2018-PSE]

[11]   Mahato A. Climate change and its impact on agriculture. Int J Sci Res 2014; 4: 1-6.

[12]   Dhawan V. Water and agriculture in India. Background paper for the South Asia expert panel during the Global Forum for Food and Agriculture. Vol. 28. 2017.

[13]   Li H, Li X, Zhang D, Liu H, Guan K. Effects of drought stress on the seed germination and early seedling growth of the endemic desert plant Eremosparton songoricum (Fabaceae). EXCLI J 2013; 12: 89-101.
[PMID: 26417219]

[14]   Okçu G, Mehmet D, Mehmet A. Effects of salt and drought stresses on germination and seedling growth of pea (Pisum sativum L.). Turkish Agric Forestry 2005; 29: 237-42.

[15]   Zeid IM, Shedeed ZA. Response of alfalfa to putrescine treatment under drought stress. Biol Plant 2006; 50(4): 635-40.
[http://dx.doi.org/10.1007/s10535-006-0099-9]

[16]   Farooq M, Wahid A, Kobayashi NS, Fujita DB, Basra SM. Plant drought stress: effects, mechanisms and management.Sustainable agriculture. Dordrecht: Springer 2009; pp. 153-88.
[http://dx.doi.org/10.1007/978-90-481-2666-8_12]

[17]   Choi WG, Toyota M, Kim SH, Hilleary R, Gilroy S. Salt stress-induced Ca^{2+} waves are associated with rapid, long-distance root-to-shoot signaling in plants. Proc Natl Acad Sci USA 2014; 111(17): 6497-502.
[http://dx.doi.org/10.1073/pnas.1319955111] [PMID: 24706854]

[18]   Harmon AC, Gribskov M, Harper JF. CDPKs – a kinase for every Ca^{2+} signal? Trends Plant Sci 2000; 5(4): 154-9.
[http://dx.doi.org/10.1016/S1360-1385(00)01577-6] [PMID: 10740296]

[19]   Weinl S, Kudla J. The CBL–CIPK Ca^{2+} decoding signaling network: function and perspectives. New Phytol 2009; 184(3): 517-28.
[http://dx.doi.org/10.1111/j.1469-8137.2009.02938.x] [PMID: 19860013]

[20]   Yoo JH, Park CY, Kim JC, *et al.* Direct interaction of a divergent CaM isoform and the transcription factor, MYB2, enhances salt tolerance in arabidopsis. J Biol Chem 2005; 280(5): 3697-706.
[http://dx.doi.org/10.1074/jbc.M408237200] [PMID: 15569682]

[21]   Weng H, Yoo CY, Gosney MJ, Hasegawa PM, Mickelbart MV. Poplar GTL1 is a Ca^{2+}/calmodulin-binding transcription factor that functions in plant water use efficiency and drought tolerance. PLoS One 2012; 7(3): e32925.
[http://dx.doi.org/10.1371/journal.pone.0032925] [PMID: 22396800]

[22]   Pandey N, Ranjan A, Pant P, *et al.* CAMTA 1 regulates drought responses in Arabidopsis thaliana. BMC Genomics 2013; 14(1): 216.
[http://dx.doi.org/10.1186/1471-2164-14-216] [PMID: 23547968]

[23]   Singh A, Kumar A, Yadav S, Singh IK. Reactive oxygen species-mediated signaling during abiotic stress. Plant Gene 2019; 18: 100173.
[http://dx.doi.org/10.1016/j.plgene.2019.100173]

[24]   Kiegle E, Moore CA, Haseloff J, Tester MA, Knight MR. Cell-type-specific calcium responses to drought, salt and cold in the *Arabidopsis* root. Plant J 2000; 23(2): 267-78.
[http://dx.doi.org/10.1046/j.1365-313x.2000.00786.x] [PMID: 10929120]

[25]   Donaldson L, Ludidi N, Knight MR, Gehring C, Denby K. Salt and osmotic stress cause rapid increases in *Arabidopsis thaliana* cGMP levels. FEBS Lett 2004; 569(1-3): 317-20.
[http://dx.doi.org/10.1016/j.febslet.2004.06.016] [PMID: 15225654]

[26]   Maathuis FJM, Sanders D. Sodium uptake in Arabidopsis roots is regulated by cyclic nucleotides. Plant Physiol 2001; 127(4): 1617-25.
[http://dx.doi.org/10.1104/pp.010502] [PMID: 11743106]

[27]   Maathuis FJM. cGMP modulates gene transcription and cation transport in Arabidopsis roots. Plant J 2006; 45(5): 700-11.
[http://dx.doi.org/10.1111/j.1365-313X.2005.02616.x] [PMID: 16460505]

[28]   Francisco R, Pilar F, Josefa MNV. Effects of Ca^{2+}, K^+ and cGMP on Na^+ uptake in pepper plants. Plant Sci 2003; 65: 1043-9.

[29]   Zhu JK. Genetic analysis of plant salt tolerance using Arabidopsis. Plant Physiol 2000; 124(3): 941-8.
[http://dx.doi.org/10.1104/pp.124.3.941] [PMID: 11080272]

[30]   Duan L, Dietrich D, Ng CH, *et al.* Endodermal ABA signaling promotes lateral root quiescence during salt stress in Arabidopsis seedlings. Plant Cell 2013; 25(1): 324-41.
[http://dx.doi.org/10.1105/tpc.112.107227] [PMID: 23341337]

[31]   Zhu JK. Salt and drought stress signal transduction in plants. Annu Rev Plant Biol 2002; 53(1): 247-73.
[http://dx.doi.org/10.1146/annurev.arplant.53.091401.143329] [PMID: 12221975]

[32]   Cheng WH, Endo A, Zhou L, *et al.* A unique short-chain dehydrogenase/reductase in Arabidopsis glucose signaling and abscisic acid biosynthesis and functions. Plant Cell 2002; 14(11): 2723-43.
[http://dx.doi.org/10.1105/tpc.006494] [PMID: 12417697]

[33]   Munnik T. Phosphatidic acid: an emerging plant lipid second messenger. Trends Plant Sci 2001; 6(5): 227-33.
[http://dx.doi.org/10.1016/S1360-1385(01)01918-5] [PMID: 11335176]

[34]   Sewelam N, Oshima Y, Mitsuda N, Ohme-Takagi M. A step towards understanding plant responses to multiple environmental stresses: a genome-wide study. Plant Cell Environ 2014; 37(9): 2024-35.
[http://dx.doi.org/10.1111/pce.12274] [PMID: 24417440]

[35]   Sanders D. Plant biology: The salty tale of Arabidopsis. Curr Biol 2000; 10(13): R486-8.
[http://dx.doi.org/10.1016/S0960-9822(00)00554-6] [PMID: 10898972]

[36]   Shi H, Zhu JK. Regulation of expression of the vacuolar Na^+/H^+ antiporter gene AtNHX1 by salt stress and abscisic acid. Plant Mol Biol 2002; 50(3): 543-50.
[http://dx.doi.org/10.1023/A:1019859319617] [PMID: 12369629]

[37]   Liu J, Ishitani M, Halfter U, Kim CS, Zhu JK. The *Arabidopsis thaliana* SOS2 gene encodes a protein kinase that is required for salt tolerance. Proc Natl Acad Sci USA 2000; 97(7): 3730-4.
[http://dx.doi.org/10.1073/pnas.97.7.3730] [PMID: 10725382]

[38]   Ishitani M, Liu J, Halfter U, Kim CS, Shi W, Zhu JK. SOS3 function in plant salt tolerance requires N-myristoylation and calcium binding. Plant Cell 2000; 12(9): 1667-77.
[http://dx.doi.org/10.1105/tpc.12.9.1667] [PMID: 11006339]

[39]   Yang Q, Chen ZZ, Zhou XF, *et al.* Overexpression of SOS (Salt Overly Sensitive) genes increases salt tolerance in transgenic Arabidopsis. Mol Plant 2009; 2(1): 22-31.

[http://dx.doi.org/10.1093/mp/ssn058] [PMID: 19529826]

[40]   Martínez-Atienza J, Jiang X, Garciadeblas B, *et al.* Conservation of the salt overly sensitive pathway in rice. Plant Physiol 2007; 143(2): 1001-12.
[http://dx.doi.org/10.1104/pp.106.092635] [PMID: 17142477]

[41]   Blumwald E, Poole RJ. Na^+/H^+ antiport in isolated tonoplast vesicles from storage tissue of Beta vulgaris. Plant Physiol 1985; 78(1): 163-7.
[http://dx.doi.org/10.1104/pp.78.1.163] [PMID: 16664191]

[42]   Blumwald E, Poole RJ. Salt tolerance in suspension cultures of sugar beet : induction of na/h antiport activity at the tonoplast by growth in salt. Plant Physiol 1987; 83(4): 884-7.
[http://dx.doi.org/10.1104/pp.83.4.884] [PMID: 16665356]

[43]   Apse MP, Aharon GS, Snedden WA, Blumwald E. Salt tolerance conferred by overexpression of a vacuolar Na^+/H^+ antiport in Arabidopsis. Science 1999; 285(5431): 1256-8.
[http://dx.doi.org/10.1126/science.285.5431.1256] [PMID: 10455050]

[44]   Zhang HX, Blumwald E. Transgenic salt-tolerant tomato plants accumulate salt in foliage but not in fruit. Nat Biotechnol 2001; 19(8): 765-8.
[http://dx.doi.org/10.1038/90824] [PMID: 11479571]

[45]   Qi Z, Spalding EP. Protection of plasma membrane K^+ transport by the salt overly sensitive1 Na^+-H^+ antiporter during salinity stress. Plant Physiol 2004; 136(1): 2548-55.
[http://dx.doi.org/10.1104/pp.104.049213] [PMID: 15347782]

[46]   Tahir MA, Aziz T, Farooq M, Sarwar G. Silicon-induced changes in growth, ionic composition, water relations, chlorophyll contents and membrane permeability in two salt-stressed wheat genotypes. Arch Agron Soil Sci 2012; 58(3): 247-56.
[http://dx.doi.org/10.1080/03650340.2010.518959]

[47]   Munns R. Genes and salt tolerance: bringing them together. New Phytol 2005; 167(3): 645-63.
[http://dx.doi.org/10.1111/j.1469-8137.2005.01487.x] [PMID: 16101905]

[48]   Szabados L, Savouré A. Proline: a multifunctional amino acid. Trends Plant Sci 2010; 15(2): 89-97.
[http://dx.doi.org/10.1016/j.tplants.2009.11.009] [PMID: 20036181]

[49]   Not Available NA, Mohanty P, Matysik J. Effect of proline on the production of singlet oxygen. Amino Acids 2001; 21(2): 195-200.
[http://dx.doi.org/10.1007/s007260170026] [PMID: 11665815]

[50]   Rehman AU, Bashir F, Ayaydin F, Kóta Z, Páli T, Vass I. Proline is a quencher of singlet oxygen and superoxide both in *in vitro* systems and isolated thylakoids. Physiol Plant 2021; 172(1): 7-18.
[http://dx.doi.org/10.1111/ppl.13265] [PMID: 33161571]

[51]   Ben Ahmed C, Ben Rouina B, Sensoy S, Boukhriss M, Ben Abdullah F. Exogenous proline effects on photosynthetic performance and antioxidant defense system of young olive tree. J Agric Food Chem 2010; 58(7): 4216-22.
[http://dx.doi.org/10.1021/jf9041479] [PMID: 20210359]

[52]   Okuma E, Soeda K, Tada M, Murata Y. Exogenous proline mitigates the inhibition of growth of *Nicotiana tabacum* cultured cells under saline conditions. Soil Sci Plant Nutr 2000; 46(1): 257-63.
[http://dx.doi.org/10.1080/00380768.2000.10408781]

[53]   Deivanai S, Xavier R, Vinod V, Timalata K, Lim OF. Role of exogenous proline in ameliorating salt stress at early stage in two rice cultivars. J Stress Physiol Biochem 2011; 7: 157-74.

[54]   Jain M, Mathur G, Koul S, Sarin N. Ameliorative effects of proline on salt stress-induced lipid peroxidation in cell lines of groundnut (*Arachis hypogaea L.*). Plant Cell Rep 2001; 20(5): 463-8.
[http://dx.doi.org/10.1007/s002990100353]

[55]   Wang Y, Nii N. Changes in chlorophyll, ribulose bisphosphate carboxylase-oxygenase, glycine betaine content, photosynthesis and transpiration in *Amaranthus tricolor* leaves during salt stress. J

Hortic Sci Biotechnol 2000; 75(6): 623-7.
[http://dx.doi.org/10.1080/14620316.2000.11511297]

[56]　Gupta B, Huang B. Mechanism of salinity tolerance in plants: physiological, biochemical, and molecular characterization. Int J Genomics 2014; 2014: 1-18.
[http://dx.doi.org/10.1155/2014/701596] [PMID: 24804192]

[57]　Ahmad R, Lim CJ, Kwon S-Y. Glycine betaine: a versatile compound with great potential for gene pyramiding to improve crop plant performance against environmental stresses. Plant Biotechnol Rep 2013; 7(1): 49-57.
[http://dx.doi.org/10.1007/s11816-012-0266-8]

[58]　Rahman S, Miyake H, Takeoka Y. Effects of exogenous glycinebetaine on growth and ultrastructure of salt-stressed rice seedlings (*Oryza sativa L.*). Plant Prod Sci 2002; 5(1): 33-44.
[http://dx.doi.org/10.1626/pps.5.33]

[59]　Cha-um S, Thapanee S, Chalermpol K. Glycine betaine alleviates water deficit stress in indica rice using proline accumulation, photosynthetic efficiencies, growth performances and yield attributes. Aust J Crop Sci 2013; 7: 213-8.

[60]　Williamson JD, Jennings DB, Guo WW, Pharr DM, Ehrenshaft M. Sugar alcohols, salt stress, and fungal resistance: polyols—multifunctional plant protection? J Am Soc Hortic Sci 2002; 127(4): 467-73.
[http://dx.doi.org/10.21273/JASHS.127.4.467]

[61]　Kerepesi I, Galiba G. Osmotic and salt stress-induced alteration in soluble carbohydrate content in wheat seedlings. Crop Sci 2000; 40(2): 482-7.
[http://dx.doi.org/10.2135/cropsci2000.402482x]

[62]　Parida AK, Das AB, Sanada Y, Mohanty P. Effects of salinity on biochemical components of the mangrove, Aegiceras corniculatum. Aquat Bot 2004; 80(2): 77-87.
[http://dx.doi.org/10.1016/j.aquabot.2004.07.005]

[63]　Raghavendra AS, Gonugunta VK, Christmann A, Grill E. ABA perception and signalling. Trends Plant Sci 2010; 15(7): 395-401.
[http://dx.doi.org/10.1016/j.tplants.2010.04.006] [PMID: 20493758]

[64]　Gurmani AR, Bano A, Khan SU, Din J, Zhang JL. Alleviation of salt stress by seed treatment with abscisic acid (ABA), 6-benzylaminopurine (BA) and chlormequat chloride (CCC) optimizes ion and organic matter accumulation and increases yield of rice (Oryza sativa L.). Aust J Crop Sci 2011; 5: 1278-85.

[65]　Xiong L, Schumaker KS, Zhu JK. Cell signaling during cold, drought, and salt stress. Plant Cell 2002; 14(Suppl) (Suppl. 1): S165-83.
[http://dx.doi.org/10.1105/tpc.000596] [PMID: 12045276]

[66]　Du H, Wang N, Cui F, Li X, Xiao J, Xiong L. Characterization of the β-carotene hydroxylase gene DSM2 conferring drought and oxidative stress resistance by increasing xanthophylls and abscisic acid synthesis in rice. Plant Physiol 2010; 154(3): 1304-18.
[http://dx.doi.org/10.1104/pp.110.163741] [PMID: 20852032]

[67]　Hadiarto T, Tran LSP. Progress studies of drought-responsive genes in rice. Plant Cell Rep 2011; 30(3): 297-310.
[http://dx.doi.org/10.1007/s00299-010-0956-z] [PMID: 21132431]

[68]　Zolla G, Heimer YM, Barak S. Mild salinity stimulates a stress-induced morphogenic response in Arabidopsis thaliana roots. J Exp Bot 2010; 61(1): 211-24.
[http://dx.doi.org/10.1093/jxb/erp290] [PMID: 19783843]

[69]　Wang Y, Li K, Li X. Auxin redistribution modulates plastic development of root system architecture under salt stress in Arabidopsis thaliana. J Plant Physiol 2009; 166(15): 1637-45.
[http://dx.doi.org/10.1016/j.jplph.2009.04.009] [PMID: 19457582]

[70] Jung JH, Park CM. Auxin modulation of salt stress signaling in Arabidopsis seed germination. Plant Signal Behav 2011; 6(8): 1198-200.
[http://dx.doi.org/10.4161/psb.6.8.15792] [PMID: 21757997]

[71] Dunlap JR, Binzel ML. NaCl reduces indole-3-acetic acid levels in the roots of tomato plants independent of stress-induced abscisic acid. Plant Physiol 1996; 112(1): 379-84.
[http://dx.doi.org/10.1104/pp.112.1.379] [PMID: 12226396]

[72] Hadiarto T, Tran LSP. Progress studies of drought-responsive genes in rice. Plant Cell Rep 2011; 30(3): 297-310.
[http://dx.doi.org/10.1007/s00299-010-0956-z] [PMID: 21132431]

[73] Yoon JY, Hamayun M, Lee SK, Lee IJ. Methyl jasmonate alleviated salinity stress in soybean. J Crop Sci Biotechnol 2009; 12(2): 63-8.
[http://dx.doi.org/10.1007/s12892-009-0060-5]

[74] Kang DJ, Seo YJ, Lee JD, *et al.* Jasmonic acid differentially affects growth, ion uptake and abscisic acid concentration in salt-tolerant and salt-sensitive rice cultivars. J Agron Crop Sci 2005; 191(4): 273-82.
[http://dx.doi.org/10.1111/j.1439-037X.2005.00153.x]

[75] Javid MG, Sorooshzadeh A, Moradi F, Modarres SSA, Allahdadi I. The role of phytohormones in alleviating salt stress in crop plants. Aust J Crop Sci 2011; 5: 726-34.

[76] Cao WH, Liu J, He XJ, *et al.* Modulation of ethylene responses affects plant salt-stress responses. Plant Physiol 2007; 143(2): 707-19.
[http://dx.doi.org/10.1104/pp.106.094292] [PMID: 17189334]

[77] Wang Y, Wang T, Li K, Li X. Genetic analysis of involvement of ETR1 in plant response to salt and osmotic stress. Plant Growth Regul 2008; 54(3): 261-9.
[http://dx.doi.org/10.1007/s10725-007-9249-0]

[78] Achard P, Cheng H, De Grauwe L, *et al.* Integration of plant responses to environmentally activated phytohormonal signals. Science 2006; 311(5757): 91-4.
[http://dx.doi.org/10.1126/science.1118642] [PMID: 16400150]

[79] He XJ, Mu RL, Cao WH, Zhang ZG, Zhang JS, Chen SY. AtNAC2, a transcription factor downstream of ethylene and auxin signaling pathways, is involved in salt stress response and lateral root development. Plant J 2005; 44(6): 903-16.
[http://dx.doi.org/10.1111/j.1365-313X.2005.02575.x] [PMID: 16359384]

[80] Cao WH, Liu J, Zhou QY, *et al.* Expression of tobacco ethylene receptor NTHK1 alters plant responses to salt stress. Plant Cell Environ 2006; 29(7): 1210-9.
[http://dx.doi.org/10.1111/j.1365-3040.2006.01501.x] [PMID: 17080944]

[81] Chaves MM, Flexas J, Pinheiro C. Photosynthesis under drought and salt stress: regulation mechanisms from whole plant to cell. Ann Bot (Lond) 2009; 103(4): 551-60.
[http://dx.doi.org/10.1093/aob/mcn125] [PMID: 18662937]

[82] Biswal B, Joshi PN, Raval MK, Biswal UC. Photosynthesis, a global sensor of environmental stress in green plants: stress signalling and adaptation. Curr Sci 2011; 47-56.

[83] Silva EN, Ribeiro RV, Ferreira-Silva SL, Viégas RA, Silveira JAG. Salt stress induced damages on the photosynthesis of physic nut young plants. Sci Agric 2011; 68(1): 62-8.
[http://dx.doi.org/10.1590/S0103-90162011000100010]

[84] Jajoo A. Changes in photosystem II in response to salt stress. Ecophysiology and responses of plants under salt stress. New York: Springer 2013; pp. 149-68.
[http://dx.doi.org/10.1007/978-1-4614-4747-4_5]

[85] Noctor G, De Paepe R, Foyer CH. Mitochondrial redox biology and homeostasis in plants. Trends Plant Sci 2007; 12(3): 125-34.

[http://dx.doi.org/10.1016/j.tplants.2007.01.005] [PMID: 17293156]

[86]   Miller G, Suzuki N, Ciftci-Yilmaz S, Mittler R. Reactive oxygen species homeostasis and signalling during drought and salinity stresses. Plant Cell Environ 2010; 33(4): 453-67.
[http://dx.doi.org/10.1111/j.1365-3040.2009.02041.x] [PMID: 19712065]

[87]   Allakhverdiev SI, Nishiyama Y, Miyairi S, *et al.* Salt stress inhibits the repair of photodamaged photosystem II by suppressing the transcription and translation of psbA genes in synechocystis. Plant Physiol 2002; 130(3): 1443-53.
[http://dx.doi.org/10.1104/pp.011114] [PMID: 12428009]

[88]   Foyer CH, Noctor G. Redox sensing and signalling associated with reactive oxygen in chloroplasts, peroxisomes and mitochondria. Physiol Plant 2003; 119(3): 355-64.
[http://dx.doi.org/10.1034/j.1399-3054.2003.00223.x]

[89]   Bafeel SO, Ibrahim MM. Antioxidants and accumulation of α-tocopherol induce chilling tolerance in Medicago sativa. Int J Agric Biol 2008; 10: 593-08.

[90]   Ouyang S, He S, Liu P, Zhang W, Zhang J, Chen S. The role of tocopherol cyclase in salt stress tolerance of rice (*Oryza sativa*). Sci China Life Sci 2011; 54(2): 181-8.
[http://dx.doi.org/10.1007/s11427-011-4138-1] [PMID: 21318489]

[91]   Gomathi R, Rakkiyapan P. Comparative lipid peroxidation, leaf membrane thermostability, and antioxidant system in four sugarcane genotypes differing in salt tolerance. Int J Plant Physiol Biochem 2011; 3: 67-74.

[92]   Rice-Evans C, Miller N, Paganga G. Antioxidant properties of phenolic compounds. Trends Plant Sci 1997; 2(4): 152-9.
[http://dx.doi.org/10.1016/S1360-1385(97)01018-2]

[93]   Valcarcel J, Reilly K, Gaffney M, O'Brien NM. Antioxidant activity, total phenolic and total flavonoid content in sixty varieties of potato (*Solanum tuberosum L.*) grown in Ireland. Potato Res 2015; 58(3): 221-44.
[http://dx.doi.org/10.1007/s11540-015-9299-z]

[94]   Gupta AS, Webb RP, Holaday AS, Allen RD. Overexpression of superoxide dismutase protects plants from oxidative stress (induction of ascorbate peroxidase in superoxide dismutase-overexpressing plants). Plant Physiol 1993; 103(4): 1067-73.
[http://dx.doi.org/10.1104/pp.103.4.1067] [PMID: 12232001]

[95]   Mandhania S, Madan S, Sawhney V. Antioxidant defense mechanism under salt stress in wheat seedlings. Biol Plant 2006; 50(2): 227-31.
[http://dx.doi.org/10.1007/s10535-006-0011-7]

[96]   Noreen Z, Ashraf M. Assessment of variation in antioxidative defense system in salt-treated pea (*Pisum sativum*) cultivars and its putative use as salinity tolerance markers. J Plant Physiol 2009; 166(16): 1764-74.
[http://dx.doi.org/10.1016/j.jplph.2009.05.005] [PMID: 19540015]

[97]   Wang Y, Ying Y, Chen J, Wang X. Transgenic Arabidopsis overexpressing Mn-SOD enhanced salt-tolerance. Plant Sci 2004; 167(4): 671-7.
[http://dx.doi.org/10.1016/j.plantsci.2004.03.032]

[98]   Zhao MG, Chen L, Zhang LL, Zhang WH. Nitric reductase-dependent nitric oxide production is involved in cold acclimation and freezing tolerance in Arabidopsis. Plant Physiol 2009; 151(2): 755-67.
[http://dx.doi.org/10.1104/pp.109.140996] [PMID: 19710235]

[99]   Bajguz A. Nitric oxide: role in plants under abiotic stress.Physiological mechanisms and adaptation strategies in plants under changing environment. New York: Springer 2014; pp. 137-59.
[http://dx.doi.org/10.1007/978-1-4614-8600-8_5]

[100] Crawford NM. Mechanisms for nitric oxide synthesis in plants. J Exp Bot 2006; 57(3): 471-8.

[http://dx.doi.org/10.1093/jxb/erj050] [PMID: 16356941]

[101] Zhang F, Wang Y, Yang Y, Wu H, Wang D, Liu J. Involvement of hydrogen peroxide and nitric oxide in salt resistance in the calluses from Populus euphratica. Plant Cell Environ 2007; 30(7): 775-85.
[http://dx.doi.org/10.1111/j.1365-3040.2007.01667.x] [PMID: 17547650]

[102] Ha HC, Sirisoma NS, Kuppusamy P, Zweier JL, Woster PM, Casero RA Jr. The natural polyamine spermine functions directly as a free radical scavenger. Proc Natl Acad Sci USA 1998; 95(19): 11140-5.
[http://dx.doi.org/10.1073/pnas.95.19.11140] [PMID: 9736703]

[103] Baniasadi F, Saffari VR, Maghsoudi Moud AA. Physiological and growth responses of Calendula officinalis L. plants to the interaction effects of polyamines and salt stress. Sci Hortic (Amsterdam) 2018; 234: 312-7.
[http://dx.doi.org/10.1016/j.scienta.2018.02.069]

[104] Knott JM, Römer P, Sumper M. Putative spermine synthases from *Thalassiosira pseudonana* and *Arabidopsis thaliana* synthesize thermospermine rather than spermine. FEBS Lett 2007; 581(16): 3081-6.
[http://dx.doi.org/10.1016/j.febslet.2007.05.074] [PMID: 17560575]

[105] Minocha R, Majumdar R, Minocha SC. Polyamines and abiotic stress in plants: a complex relationship1. Front Plant Sci 2014; 5: 175.
[http://dx.doi.org/10.3389/fpls.2014.00175] [PMID: 24847338]

[106] Lutts S, Hausman JF, Quinet M, Lefèvre I. Polyamines and their roles in the alleviation of ion toxicities in plants. Ecophysiology and responses of plants under salt stress. New York: Springer 2013; pp. 315-53.
[http://dx.doi.org/10.1007/978-1-4614-4747-4_12]

[107] Fuell C, Elliott KA, Hanfrey CC, Franceschetti M, Michael AJ. Polyamine biosynthetic diversity in plants and algae. Plant Physiol Biochem 2010; 48(7): 513-20.
[http://dx.doi.org/10.1016/j.plaphy.2010.02.008] [PMID: 20227886]

[108] Tiburcio AF, Altabella T, Bitrián M, Alcázar R. The roles of polyamines during the lifespan of plants: from development to stress. Planta 2014; 240(1): 1-18.
[http://dx.doi.org/10.1007/s00425-014-2055-9] [PMID: 24659098]

[109] Yang X, Lu M, Wang Y, Wang Y, Liu Z, Chen S. Response Mechanism of Plants to Drought Stress. Horticulturae 2021; 7(3): 50.
[http://dx.doi.org/10.3390/horticulturae7030050]

[110] Chazen O, Neumann PM. Hydraulic Signals from the Roots and Rapid Cell-Wall Hardening in Growing Maize (Zea mays L.) Leaves Are Primary Responses to Polyethylene Glycol-Induced Water Deficits. Plant Physiol 1994; 104(4): 1385-92.
[http://dx.doi.org/10.1104/pp.104.4.1385] [PMID: 12232175]

[111] Fromm J, Fei H. Electrical signaling and gas exchange in maize plants of drying soil. Plant Sci 1998; 132(2): 203-13.
[http://dx.doi.org/10.1016/S0168-9452(98)00010-7]

[112] Wang WH, Yi XQ, Han AD, *et al.* Calcium-sensing receptor regulates stomatal closure through hydrogen peroxide and nitric oxide in response to extracellular calcium in Arabidopsis. J Exp Bot 2012; 63(1): 177-90.
[http://dx.doi.org/10.1093/jxb/err259] [PMID: 21940718]

[113] Case R, Eisner D, Gurney A, Jones O, Muallem S, Verkhratsky A. Evolution of calcium homeostasis: From birth of the first cell to an omnipresent signalling system. Cell Calcium 2007; 42(4-5): 345-50.
[http://dx.doi.org/10.1016/j.ceca.2007.05.001] [PMID: 17574670]

[114] Miller G, Shulaev V, Mittler R. Reactive oxygen signaling and abiotic stress. Physiol Plant 2008; 133(3): 481-9.

[http://dx.doi.org/10.1111/j.1399-3054.2008.01090.x] [PMID: 18346071]

[115] Yan J, Tsuichihara N, Etoh T, Iwai S. Reactive oxygen species and nitric oxide are involved in ABA inhibition of stomatal opening. Plant Cell Environ 2007; 30(10): 1320-5.
[http://dx.doi.org/10.1111/j.1365-3040.2007.01711.x] [PMID: 17727421]

[116] Hossain MA, Bhattacharjee S, Armin SM, *et al.* Hydrogen peroxide priming modulates abiotic oxidative stress tolerance: insights from ROS detoxification and scavenging. Front Plant Sci 2015; 6: 420.
[http://dx.doi.org/10.3389/fpls.2015.00420] [PMID: 26136756]

[117] Tanveer M, Shahzad B, Sharma A, Khan EA. 24-Epibrassinolide application in plants: An implication for improving drought stress tolerance in plants. Plant Physiol Biochem 2019; 135: 295-303.
[http://dx.doi.org/10.1016/j.plaphy.2018.12.013] [PMID: 30599306]

[118] Umezawa T, Hirayama T, Kuromori T, Shinozaki K. The regulatory networks of plant responses to abscisic acid. Adv Bot Res 2011; 57: 201-48.
[http://dx.doi.org/10.1016/B978-0-12-387692-8.00006-0]

[119] Kuromori T, Mizoi J, Umezawa T, Yamaguchi-Shinozaki K, Shinozaki K. Stress Signaling Networks: Drought Stress. Mol Biol 2013; 1-23.

[120] Park S-Y, Fung P, Nishimura N, *et al.* Abscisic acid inhibits PP2Cs *via* the PYR/PYL family of ABA-binding START proteins. Science 2009; 324(5930): 1068-71.
[http://dx.doi.org/10.1126/science.1173041] [PMID: 19407142]

[121] Abe H, Urao T, Ito T, Seki M, Shinozaki K, Yamaguchi-Shinozaki K. Arabidopsis AtMYC2 (bHLH) and AtMYB2 (MYB) function as transcriptional activators in abscisic acid signaling. Plant Cell 2003; 15(1): 63-78.
[http://dx.doi.org/10.1105/tpc.006130] [PMID: 12509522]

[122] Agarwal PK, Agarwal P, Reddy MK, Sopory SK. Role of DREB transcription factors in abiotic and biotic stress tolerance in plants. Plant Cell Rep 2006; 25(12): 1263-74.
[http://dx.doi.org/10.1007/s00299-006-0204-8] [PMID: 16858552]

[123] Okamuro JK, Caster B, Villarroel R, Van Montagu M, Jofuku KD. The AP2 domain of *APETALA2* defines a large new family of DNA binding proteins in *Arabidopsis*. Proc Natl Acad Sci USA 1997; 94(13): 7076-81.
[http://dx.doi.org/10.1073/pnas.94.13.7076] [PMID: 9192694]

[124] Zhou J, Tang X, Martin GB. The Pto kinase conferring resistance to tomato bacterial speck disease interacts with proteins that bind a cis-element of pathogenesis-related genes. EMBO J 1997; 16(11): 3207-18.
[http://dx.doi.org/10.1093/emboj/16.11.3207] [PMID: 9214637]

[125] Ohme-Takagi M, Shinshi H. Ethylene-inducible DNA binding proteins that interact with an ethylene-responsive element. Plant Cell 1995; 7(2): 173-82.
[PMID: 7756828]

[126] Sasaki T, Song J, Koga-Ban Y, *et al.* Toward cataloguing all rice genes: large-scale sequencing of randomly chosen rice cDNAs from a callus cDNA library. Plant J 1994; 6(4): 615-24.
[http://dx.doi.org/10.1046/j.1365-313X.1994.6040615.x] [PMID: 7987417]

[127] Weigel D. The APETALA2 domain is related to a novel type of DNA binding domain. Plant Cell 1995; 7(4): 388-9.
[PMID: 7773013]

[128] Moose SP, Sisco PH. Glossy15, an APETALA2-like gene from maize that regulates leaf epidermal cell identity. Genes Dev 1996; 10(23): 3018-27.
[http://dx.doi.org/10.1101/gad.10.23.3018] [PMID: 8957002]

[129] Nakashima K, Kiyosue T, Yamaguchi-Shinozaki K, Shinozaki K. A nuclear gene, erd1, encoding a chloroplast-targeted Clp protease regulatory subunit homolog is not only induced by water stress but

also developmentally up-regulated during senescence in Arabidopsis thaliana. Plant J 1997; 12(4): 851-61.
[http://dx.doi.org/10.1046/j.1365-313X.1997.12040851.x] [PMID: 9375397]

[130] Cominelli E, Galbiati M, Vavasseur A, *et al.* A guard-cell-specific MYB transcription factor regulates stomatal movements and plant drought tolerance. Curr Biol 2005; 15(13): 1196-200.
[http://dx.doi.org/10.1016/j.cub.2005.05.048] [PMID: 16005291]

[131] Liang YK, Dubos C, Dodd IC, Holroyd GH, Hetherington AM, Campbell MM. AtMYB61, an R2R3-MYB transcription factor controlling stomatal aperture in Arabidopsis thaliana. Curr Biol 2005; 15(13): 1201-6.
[http://dx.doi.org/10.1016/j.cub.2005.06.041] [PMID: 16005292]

[132] Hu H, Dai M, Yao J, *et al.* Overexpressing a NAM, ATAF, and CUC (NAC) transcription factor enhances drought resistance and salt tolerance in rice. Proc Natl Acad Sci USA 2006; 103(35): 12987-92.
[http://dx.doi.org/10.1073/pnas.0604882103] [PMID: 16924117]

[133] Latif S, Shah T, Munsif F, D'Amato R. Genetic Manipulation of Drought Stress Signaling Pathways in Plants. Salt and Drought Stress Tolerance in Plants, Signaling and Communication in Plants. Springer Nature Switzerland AG 2020; pp. 367-82.
[http://dx.doi.org/10.1007/978-3-030-40277-8_15]

[134] Govind G, Harshavardhan VT, Hong CY. Phytohormone Signaling in Response to Drought.Salt and Drought Stress Tolerance in Plants, Signaling and Communication in Plants. Springer Nature Switzerland AG 2020; pp. 367-82.
[http://dx.doi.org/10.1007/978-3-030-40277-8_12]

[135] Cheng MC, Liao PM, Kuo WW, Lin TP. The Arabidopsis ETHYLENE RESPONSE FACTOR1 regulates abiotic stress-responsive gene expression by binding to different cis-acting elements in response to different stress signals. Plant Physiol 2013; 162(3): 1566-82.
[http://dx.doi.org/10.1104/pp.113.221911] [PMID: 23719892]

[136] Feussner I, Wasternack C. The lipoxygenase pathway. Annu Rev Plant Biol 2002; 53(1): 275-97.
[http://dx.doi.org/10.1146/annurev.arplant.53.100301.135248] [PMID: 12221977]

[137] Ghasemi Pirbalouti A, Sajjadi SE, Parang K. A review (research and patents) on jasmonic acid and its derivatives. Arch Pharm (Weinheim) 2014; 347(4): 229-39.
[http://dx.doi.org/10.1002/ardp.201300287] [PMID: 24470216]

[138] McConn M, Creelman RA, Bell E, Mullet JE, Browse J. Jasmonate is essential for insect defense in *Arabidopsis*. Proc Natl Acad Sci USA 1997; 94(10): 5473-7.
[http://dx.doi.org/10.1073/pnas.94.10.5473] [PMID: 11038546]

[139] Tanaka Y, Sano T, Tamaoki M, Nakajima N, Kondo N, Hasezawa S. Ethylene inhibits abscisic acid-induced stomatal closure in Arabidopsis. Plant Physiol 2005; 138(4): 2337-43.
[http://dx.doi.org/10.1104/pp.105.063503] [PMID: 16024687]

[140] Anjum SA, Xie X, Wang L, Saleem MF, Man C, Lei W. Morphological, physiological and biochemical responses of plants to drought stress. Afr J Agric Res 2011; 6: 2026-32.

[141] Zhai Q, Zhang X, Wu F, *et al.* Transcriptional mechanism of jasmonate receptor COI1-mediated delay of flowering time in Arabidopsis. Plant Cell 2015; 27(10): tpc.15.00619.
[http://dx.doi.org/10.1105/tpc.15.00619] [PMID: 26410299]

[142] Savchenko T, Kolla VA, Wang CQ, *et al.* Functional convergence of oxylipin and abscisic acid pathways controls stomatal closure in response to drought. Plant Physiol 2014; 164(3): 1151-60.
[http://dx.doi.org/10.1104/pp.113.234310] [PMID: 24429214]

[143] Hara M, Furukawa J, Sato A, Mizoguchi Tand Miura K. Abiotic stress and role of salicylic acid in plants.Abiotic stress responses in plants. Springer 2012; pp. 235-51.
[http://dx.doi.org/10.1007/978-1-4614-0634-1_13]

[144] Catinot J, Buchala A, Abou-Mansour E, Métraux JP. Salicylic acid production in response to biotic and abiotic stress depends on isochorismate in *Nicotiana benthamiana*. FEBS Lett 2008; 582(4): 473-8.
[http://dx.doi.org/10.1016/j.febslet.2007.12.039] [PMID: 18201575]

[145] Hunter LJR, Westwood JH, Heath G, *et al.* Regulation of RNA-dependent RNA polymerase 1 and isochorismate synthase gene expression in Arabidopsis. PLoS One 2013; 8(6): e66530.
[http://dx.doi.org/10.1371/journal.pone.0066530] [PMID: 23799112]

[146] Kang G, Li G, Xu W, *et al.* Proteomics reveals the effects of salicylic acid on growth and tolerance to subsequent drought stress in wheat. J Proteome Res 2012; 11(12): 6066-79.
[http://dx.doi.org/10.1021/pr300728y] [PMID: 23101459]

[147] Senaratna T, Touchell D, Bunn E, Dixon K. Acetyl salicylic acid (Aspirin) and salicylic acid induce multiple stress tolerance in bean and tomato plants. Plant Growth Regul 2000; 30(2): 157-61.
[http://dx.doi.org/10.1023/A:1006386800974]

[148] Tang J, Han Z, Chai J. Q&A: what are brassinosteroids and how do they act in plants? BMC Biol 2016; 14(1): 113.
[http://dx.doi.org/10.1186/s12915-016-0340-8] [PMID: 28007032]

[149] Tiwari S, Lata C, Chauhan PS, Prasad V, Prasad M. A functional genomic perspective on drought signaling and its crosstalk with phytohormone-mediated signaling pathways in plants. Curr Genomics 2017; 18(6): 469-82.
[http://dx.doi.org/10.2174/1389202918666170605083319] [PMID: 29204077]

[150] Nakamura A, Tochio N, Fujioka S, *et al.* Molecular actions of two synthetic brassinosteroids, iso-carbaBL and 6-deoxoBL, which cause altered physiological activities between Arabidopsis and rice. PLoS One 2017; 12(4): e0174015.
[http://dx.doi.org/10.1371/journal.pone.0174015] [PMID: 28369122]

[151] Lacerda CF, Assis Júnior JO, Lemos Filho LCA, *et al.* Morpho-physiological responses of cowpea leaves to salt stress. Braz J Plant Physiol 2006; 18(4): 455-65.
[http://dx.doi.org/10.1590/S1677-04202006000400003]

[152] Bacha H, Tekaya M, Drine S, *et al.* Impact of salt stress on morpho-physiological and biochemical parameters of Solanum lycopersicum cv. Microtom leaves. S Afr J Bot 2017; 108: 364-9.
[http://dx.doi.org/10.1016/j.sajb.2016.08.018]

[153] Soussi M, Ocana A, Lluch C. Effects of salt stress on growth, photosynthesis and nitrogen fixation in chick-pea (Cicer arietinum L.). J Exp Bot 1998; 49(325): 1329-37.
[http://dx.doi.org/10.1093/jxb/49.325.1329]

[154] Ghaderi N, Aliakbar M, Adel S. Change in antioxidant enzymes activity and some morpho-physiological characteristics of strawberry under long-term salt stress. Physiol & Mol Bio Plants 2018 2018; 24: 833-43.

[155] Asghari R, Rahim A. Salinity Stress and its impact on Morpho-Physiological Characteristics of Aloe Vera. Pertanika, J Trop Agric Sci 2018; 41: 1.

[156] Moud AM, Kobra M. Salt stress effects on respiration and growth of germinated seeds of different wheat (*Triticum aestivum L.*) cultivars. World J Agric Sci 2008; 4: 351-8.

[157] Shaheen HL. Morpho-physiological responses of cotton (*Gossypium hirsutum*) to salt stress. Int J Agric Biol 2012; 14: 6.

[158] Azevedo Neto AD, Prisco JT, Enéas-Filho J, *et al.* Effects of salt stress on plant growth, stomatal response and solute accumulation of different maize genotypes. Braz J Plant Physiol 2004; 16(1): 31-8.
[http://dx.doi.org/10.1590/S1677-04202004000100005]

[159] Punia H, Tokas J, Mor VS, *et al.* Deciphering reserve mobilization, antioxidant potential, and

expression analysis of starch synthesis in sorghum seedlings under salt stress. Plants 2021; 10(11): 2463.
[http://dx.doi.org/10.3390/plants10112463] [PMID: 34834826]

[160] Jan SA, Zabta KS, Malik AR. Agro-morphological and physiological responses of Brassica rapa ecotypes to salt stress. Pak J Bot 2016; 48: 1379-84.

[161] Noreen S, Muhammad A. Alleviation of adverse effects of salt stress on sunflower (Helianthus annuus L.) by exogenous application of salicylic acid: growth and photosynthesis. Pak J Bot 2008; 40: 1657-63.

[162] M AH, A SJ, M B, M MH, Mohd RI, A S. Effect of salt stress on germination and early seedling growth of rice (*Oryza sativa L.*). Afr J Biotechnol 2010; 9(13): 1911-8.
[http://dx.doi.org/10.5897/AJB09.1526]

[163] Allahverdiyev T. Impact of soil water deficit on some physiological parameters of durum and bread wheat genotypes. Agriculture and Forestry. Poljopr Sumar 2016; 62: 1.
[http://dx.doi.org/10.17707/AgricultForest.62.1.16]

[164] DaMatta FM, Chaves ARM, Pinheiro HA, Ducatti C, Loureiro ME. Drought tolerance of two field-grown clones of Coffea canephora. Plant Sci 2003; 164(1): 111-7.
[http://dx.doi.org/10.1016/S0168-9452(02)00342-4]

[165] Dos S, Verissino V, Wanderley F, Ferreira VM, Cavalcante PGS, Rolim EV. Seasonal variations of photosynthesis, gas exchange, quantum efficiency of photosystem II and biochemical responses of Jatropha curcas L. grown in semi-humid and semi-arid areas subject to water stress. 2013; 41: 203-13.

[166] Hura T, Hura K, Grzesiak M, Rzepka A. Effect of long-term drought stress on leaf gas exchange and fluorescence parameters in C3 and C4 plants. Acta Physiol Plant 2007; 29(2): 103-13.
[http://dx.doi.org/10.1007/s11738-006-0013-2]

[167] Jedmowski C, Ashoub A, Momtaz O, Bruggemann W. Impact of drought, heat and their combination on chlorophyll fluorescence and yield of Wild Barley (Hordeum spontaneum). J BotArticle 2005. ID 120868 (9 pages).

[168] Liu RX, Zhou ZG, Guo WQ, Chen BL, Oosterhuis DM. Effects of N fertilization on root development and activity of water-stressed cotton (Gossypium hirsutum L.) plants. Agric Water Manage 2008; 95(11): 1261-70.
[http://dx.doi.org/10.1016/j.agwat.2008.05.002]

[169] Pan X, Lada RR, Caldwell CD, Falk KC. Water-stress and N-nutrition effects on photosynthesis and growth of Brassica carinata. Photosynthetica 2011; 49(2): 309-15.
[http://dx.doi.org/10.1007/s11099-011-0031-1]

[170] Shaw B, Thomas TH, Cooke DT. Responses of sugar beet (Beta vulgaris L.) to drought and nutrient deficiency stress. Plant Growth Regul 2002; 37(1): 77-83.
[http://dx.doi.org/10.1023/A:1020381513976]

[171] Silva EN, Ribeiro RV, Ferreira-Silva SL, Viégas RA, Silveira JAG. Comparative effects of salinity and water stress on photosynthesis, water relations and growth of Jatropha curcas plants. J Arid Environ 2010; 74(10): 1130-7.
[http://dx.doi.org/10.1016/j.jaridenv.2010.05.036]

[172] Yin CY, Berninger F, Li CY. Photosynthetic responses of Populus przewalski subjected to drought stress. Photosynthetica 2006; 44(1): 62-8.
[http://dx.doi.org/10.1007/s11099-005-0159-y]

[173] Wu FZ, Bao WK, Li FL, Wu N. Effects of water stress and nitrogen supply on leaf gas exchange and fluorescence parameters of Sophora davidii seedlings. Photosynthetica 2008; 46(1): 40-8.
[http://dx.doi.org/10.1007/s11099-008-0008-x]

CHAPTER 9

Salt Stress and its Mitigation Strategies for Enhancing Agricultural Production

Priyanka Saha[1,*], Jitendra Singh Bohra[2], Anamika Barman[1] and **Anurag Berà[2]**

[1] *Division of Agronomy, ICAR- India Agricultural Research Institute, New Delhi-110012, India*

[2] *Department of Agronomy, Institute of Agricultural Sciences, Banaras Hindu University, Varanasi, Uttar Pradesh, 221005, India*

Abstract: In agriculture, salinity has been a major limiting factor in food security. Soil salinity has been shown to limit land utilization and crop productivity. It is especially crucial to avoid such losses as the ever-increasing global population imposes a tremendous amount of pressure on human populations to produce more food and feed. Salt stress has a negative effect on the whole plant and can be seen at all phases of growth, including germination, seedling and vegetative stages. Tolerance to salt stress, on the other hand, varies with plant developmental processes and even from species and cultivars. Salinity in the agricultural system can be managed by adopting various mitigation strategies. To maintain higher productivity in salt-affected environments, salt-tolerant genotypes must be introduced, as well as precise site-specific production systems. Recent advances in genetics and biotechnology, along with traditional breeding methods, provide the potential to create transgenic cultivars that perform well under stress. Exogenous treatment of certain osmoprotectants and growth regulators, as well as nutrient management and seed rejuvenation strategies, may be beneficial for cost-effective agricultural production in saline soils

Keywords: Acid Soil, Bio-Saline Agriculture, Management Practices, Phytoremediation, Salt-Affected Soil.

1. INTRODUCTION

Since the 21st century, global water resource scarcity, increased salinization, and pollution of soil and water have become major concerns for environmentalists. Problem soils are characterized by where most plants and crops cannot be produced efficiently, and are not fertile or productive, resulting in a significant reduction in cultivated land area, yields and quality. They have substantial physical and chemical constraints on the growth of most plants, like sodicity,

* **Corresponding author Priyanka Saha :** Division of Agronomy, ICAR- India Agricultural Research Institute, New Delhi-110012, India; Email: priyankasaha9933@gmail.com

salinity, acidity, and reduced soil fertility. According to Beltran and Manzur (2005) [1], the total area under salt-affected soils, comprising sodic and saline soils, is 831 mha in the world, whereas, in India, it is about 6.7 mha. Soil acidity affects ~48 M ha of India's 142 M ha of arable land, with 25 M ha having a *p*H below 5.5 [2]. The salt-affected area continues to increase each year, hence, we need to adopt proper agronomic and soil management practices to revive the productivity of problem soils. To alleviate this, lime and/or alternate soil amendment materials, as well as other management strategies, are required. Apart from enhancing production, lime application improves fertilizer efficiency, protects the environment and raises farmers' net profit [3]. Organic amendments, such as biochar, biogas slurry, fly ash, and other organic amendments, on the other hand, have been demonstrated to be beneficial in restoring acid soils. To avoid excessive salt buildup, irrigation water needs to be provided in an excess amount of what is required for crop evaporation. When evapo-transpiration needs are low, in cold weather, and with high humidity, leaching is more efficient in preventing salt accumulation in the root zone. Other options include green manuring, mulching, and salt scraping, along with new technologies such as phytoremediation and bio-saline agriculture offer a way to grow salt-tolerant crops in adverse conditions. Biochar application reduces soil electrical conductivity and exchangeable Na percentage to an acceptable range [4]. The microbial and biological activity of sodic soils has been reported to improve when they are covered by trees, grasses, or any cultivable crops [5]. Those agronomic interventions can improve the biological, physical, and chemical properties of problem soil, which ultimately leads to an increase in crop productivity and sustainable yield. It is necessary to develop salt stress-tolerant crop varieties through breeding and plant genetic modification, but this is a time-consuming process, while agronomic interventions to relieve stress might be a more cost-effective and eco-friendly approach that could be framed in a shorter period. Hence to restore agricultural production under constrained resource conditions, it is necessary to integrate soil, water, forest, and biological resources and adopt these practices in an integrated manner that would be a greater step toward the restoration of problem soils.

2. BACKGROUND

The world population crossed 6 billion in 2000 and is expected to approach 11 billion by 2100. The world's demand for food, fiber, and bio-energy products will grow at an annual rate of 2.5% and that of developing countries at 3.7% [6] as against the annual growth of only 1% in the output. Among the total land resource of 13.5 billion ha in the world, roughly 3.03 billion ha (22%) is potentially available for cultivation, out of which only one-half (1.5 billion ha) is cultivated due to some constraints. Therefore, very limited land is available for cultivation. In comparison to the global scenario, the situation in India is grimmer with a geographical area of 329 M ha (2.3% of the total world's land), supporting about 18% human population and 15% livestock population. The amount of arable land per person is only 0.15 ha, and by 2025, that number is predicted to drop to 0.08 ha.

Owing to increase demographic pressure on the limited soil resource and unscientific practices adopted for short-term gains to meet mounting multiple demands without consideration of long-term sustained productivity, the global economy governed by urbanization, industrialization, and other developmental activities is outpacing the "land capability." As a result, grave worries about environmental and land degradation as well as a slowing rate of productivity growth, have gained prominence recently. Every year, 5-7 million hectares of arable land are lost to deterioration, and 2 billion ha of land have already undergone some form of degradation. According to estimates, soil degradation has a relative impact of 39% in Asia, 25% in Africa, 12% in South America, 11% in Europe, and 5% in Oceania. India is one of the poorest developing nations, with a high population density. The NBSS&LUP estimates that 148.9 million hectares of degraded land are affected by water erosion, followed by 13.5 million by wind erosion, 13.8 million by chemical degradation, and 11.6 million by physical degradation.

The agricultural situation in India is fast changing in response to the numerous stressors that cultivated areas are facing Fig (**1**). The agricultural sector cannot afford to wait and must act now. In the production-to-consumption chain, manage change and meet the rising and diverse needs. Nearly variable rainfall affects 7.0 million acres of agricultural land. In the country, there are various levels of salt concerns. As a result of secondary irrigation commands with salinization and lift irrigated projects, an increase in agriculture's reliance on poor seawater intrusion and brackish water aquaculture in coastal regions, as well as quality waters in semi-arid and desert regions, the problem's location is predicted to worsen shortly. India's anticipated area under salt-affected soils by 2025 is 13 million ha Table (**1**). Problem soils are the soils whose productivity is lowered due to inherent unfavorable soil conditions *viz.*, salt content and soil reaction. Through agronomic research, agronomists have played and will continue to play a crucial role in managing these lands and boosting productivity. They have developed a thorough understanding and improved contingency plans based on resource-efficient, socioeconomically viable, and ecologically sound technologies to deal with salt-affected soils in a climate variability scenario.

Numerous studies have been conducted on the characteristics of moisture retention [7, 8]. Studies on water transmission characteristics gained a lot of interest in sodic soils with varied ESP and pH. By this time, soil scientists and agronomists had discovered that cation exchange equilibrium is the most crucial element in determining how salt-affected soils react to reclamation methods. It is a perplexing situation where, on the one hand, the limited soil base must produce more to keep up with the demands of a constantly growing population, but on the other, vast areas are either losing their ability to support agriculture or are experiencing alarming productivity declines as a result of degradation at an unacceptable rate. Consequently, one of the major global problems is utilizing the productive potential of damaged lands while preventing future degradation. The potentially profitable salt-affected lands in this situation have great chances for increasing production, which is desperately required.

3. PROBLEM SOILS AND THEIR FEATURES

3.1. Acid Soil

Acid soils are base unsaturated soils that have absorbed enough H+ ions to lower the pH of the soil below 7. All critical plant nutrients' chemical solubility and precipitation are controlled by soil acidity, which affects their availability. The fertility of the soil and plant growth are both significantly impacted by soil acidity. Ca, Mg, P, B and Mo, for instance, become deficient in very acidic soils, whilst Mn and Fe may reach dangerous levels.

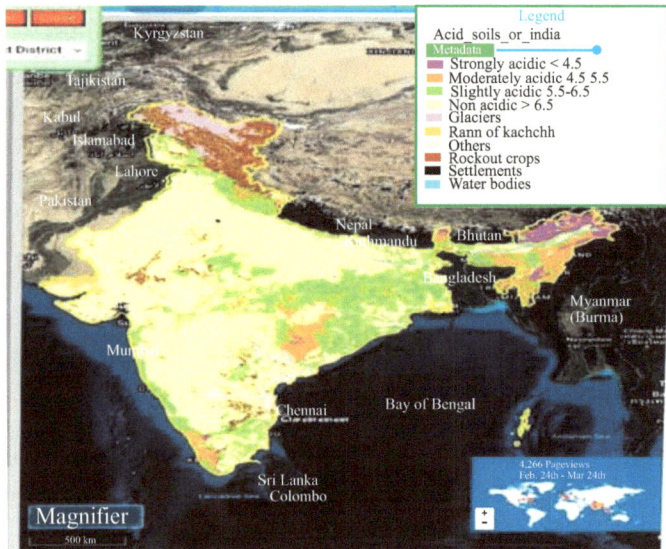

Fig. (1). Acid soil distribution in India, Source: NBSS & LUP.

3.2. Salt-affected Soils

A high amount of soluble mineral salts in the soil profile of salt-affected soils has a detrimental effect on crop performance and soil health. Salt-affected soils are mostly made up of chloride, sulfates, carbonates, and bicarbonates of calcium, magnesium, and sodium. Approximately 6.73 million acres of India's land are damaged by salt Fig (**2**).

Category	pH
Extremely acidic	**<4.5**
Very strongly acidic	**4.5-5.0**
Strongly acidic	**5.1-5.5**
Moderately acidic	**5.6-6.0**
Slightly acidic	**6.1-6.5**
Neutral	**6.6-7.3**
Slightly alkaline	**7.4-7.8**
Moderately alkaline	**7.9-8.4**
Strongly alkaline	**8.5-9.0**
Very strongly alkaline	**>9.0**

Table. (1). Acid soil distribution in India, Source: NBSS & LUP.

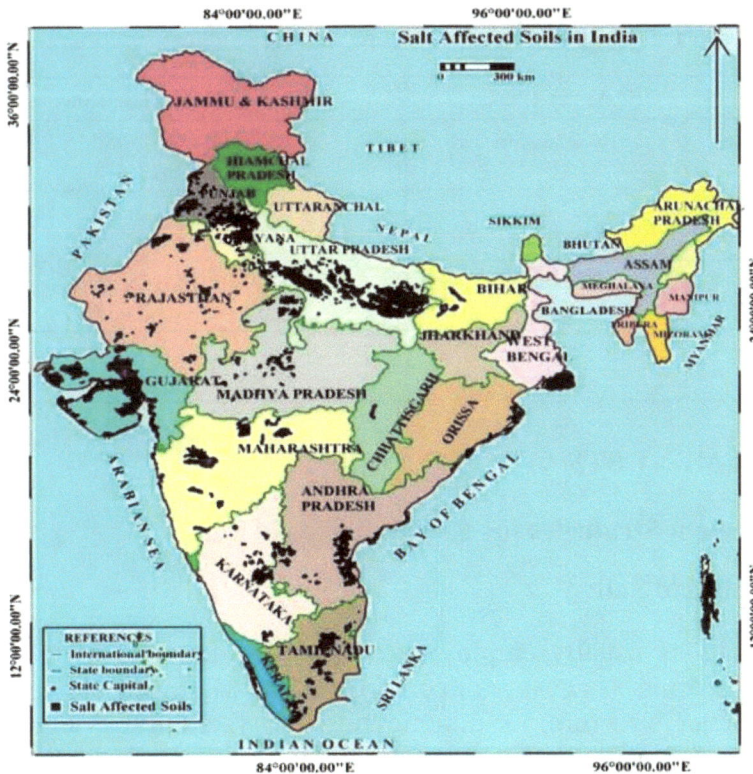

Fig. (2). Distribution of salt affected soil, Source: NBSS-LUP.

4. DIAGNOSTIC CRITERIA AND CLASSIFICATION

According to USSL [10], the demarcation between the saline and non-saline soils in terms of EC of saturation extract at 25° C is 4 dS/m. However, the adverse effect of even 2 dS/m has been observed on the growth of sensitive crops. Similarly, USSL [10] recommended pH 8.5 and ESP 15 to distinguish the sodic and non-sodic soils. Agarwal and Yadav (1956) [11] found pH 8.2 as the limiting value of ESP 15 for the salt-affected soils of the Indo-Gangatic alluvium. Many of these soils with pH 8.2-10.0 are associated with ESP 15-40 and pH>10 with ESP values ranging from 40-100.

Subsequently, based on the research conducted at CSSRI, Karnal, Abrol *et al.* (1980) [7] also suggested pH 8.2 as a better criterion for distinguishing sodic soils from non-sodic soils. The limits of values for the classification of saline, sodic and saline-sodic soils given by USSL [10], along with those suggested by the Soil Science Society of America (1987), are presented in Table **2**. The classification given by USSL is still widely used in many parts of the world. Given the recent developments, a re-look at classification is, therefore, needed.

Table 2. Classification of salt-affected soils

Class	As per USSL	Proposed by Soil Sci. Soc. America
Normal	ECe<4 dSm^{-1} ESP <15	ECe<2 dSm^{-1} SAR <13
Saline	ECe >4 dSm^{-1} ESP <15	ECe>2 dSm^{-1} SAR <13
Sodic	ECe variable ESP >15	ECe variable SAR >13
Saline-sodic	ECe >4 dSm^{-1} ESP >15	ECe>2 dSm^{-1} SAR >13

5. MANAGEMENT STRATEGIES

5.1. Management Strategies for Reclaiming Acid Soil

☐ Liming of Acid Soil:

Liming operations usually employ carbonates, oxides, and hydroxides of Ca and/or Mg. It offers a cheap source of Ca^{2+} and Mg^{2+} and enhances physical condition (better structure), symbiotic nitrogen fixation by legumes, fodder palatability, and lowers Mn^{2+} and Al^{3+} toxicity.

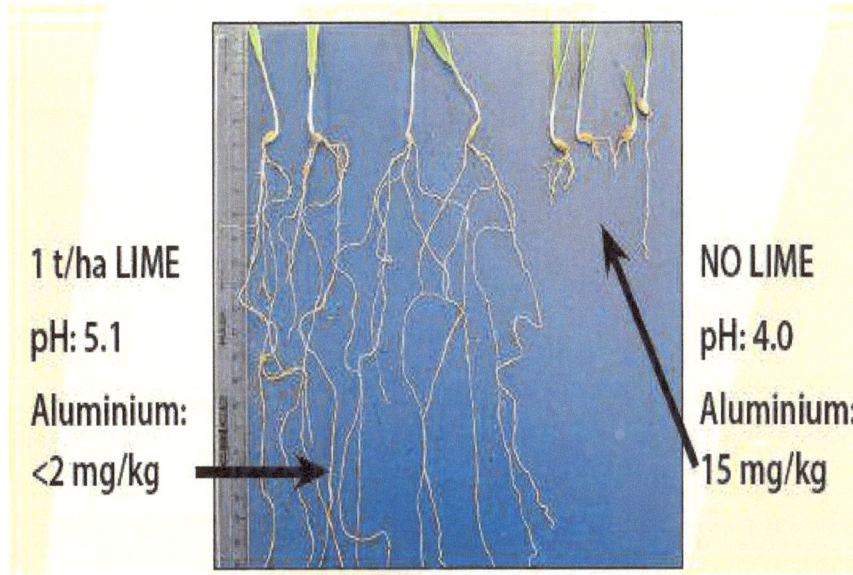

Fig. (3). Roots of barley grown in acidic subsurface soil are shortened by aluminium toxicity Source: Soil quality fact sheet, Department of Agriculture Australian Govt. website.

☐ Use of Organic Amendments

- Biogas slurry is a byproduct of the production of biogas and a natural fertilizer rich in nitrogen, phosphorus, potassium, amino acids, and other bioactive compounds.
- To transform biomass carbon into a more stable form, black carbon from biomass sources (such as wood chips, plant residues, manure, or other agricultural waste products) is created as biochar (carbon sequestration).
- As a byproduct of making paper, paper mill sludge is created, and the mill has trouble disposing of this waste. Calcium carbonate ($CaCO_3$) is the main ingredient in lime mud, and it is estimated that 0.47 m3 of lime mud is produced for every tonne of pulp [12].
- Fly ash (FA), a byproduct of coal combustion from thermal power plants, is a common soil enhancer used all over the world. FA is used to improve soil physical conditions and adjust pH. It often contains important plant micro- and macronutrients in addition to having special physicochemical qualities.

☐ **Crop Management Practices:**

○ The use of basic fertilizer can help reclaim soil acidity temporarily.
○ Selecting acid-tolerant crops helps avoid the ill effect of acid soil.

5.2. Management Strategies for Reclaiming Sodic Soil

☐ **Chemical Amendments**

Replacement of exchangeable sodium partially or wholly by soluble calcium in the root zone is a prerequisite for sodic soil reclamation. The addition of a suitable inorganic material accelerates the reclamation process. The inorganic amendments are broadly grouped into two categories:

A. Calcium salts of varying solubility, *e.g.*, gypsum, calcium chloride, ground limestone, phospho-gypsum.
B. Acid or acid-forming substances, *e.g.*, sulphuric acid, ferrous sulfate, aluminum sulfate, sulfur, lime sulfur, iron pyrite, and fly ash.

The quantity of amendment to be added depends on the severity of soil deterioration, soil texture, depth of soil to be reclaimed, kind of crop to be grown, level of improvement desired, economic considerations, *etc.* The choice of an amendment depends largely on the nature of the soil, easy availability, effectiveness, and cost, handling skill and infrastructure needed, hazard risk involved, *etc.* Yadav and Agarwal (1959) [13] suggested that about a 60% reduction in the exchangeable sodium content would constitute the near economic application of calcareous amendment like gypsum, and full replacement of exchangeable Na is not required initially. Before amendment application, flushing of soluble carbonate salts helps reduce the number of amendment requirements.

☐ **Organic Amendments**

The positive role of organic materials in facilitating the reclamation of sodic soils through the improvement of physical condition, greater mobilization of native Ca due to CO_2 evolution, the addition of many macro and micro plant nutrients, reduction in pH, and enhancement of biological activity has been known for long [13, 14]. Several organic sources include green manures, compost, FYM, crop residues such as paddy straw, industry by-products like press-mud and molasses, and weeds such as *Argemone maxicana* and water hyacinth, but their effectiveness as the sole application is much lower than that of the chemical amendments and varies greatly among themselves. *Sesbania*, as a green manure crop, has the advantage of relatively higher tolerance to soil sodicity, greater biomass production and faster decomposition, narrow C/N ratio, and appreciable N-fixing capacity. *Argemone maxicana* was recommended by Banthra Farm, but its use as an amendment on the field scale has not been popular. Due to limited availability, the potential for other industrial uses, handling care, *etc.*, large-scale adoption of molasses has been highly constrained. Press-mud obtained from the sulphation process proved superior to that obtained from the carbonation process.

Thus, there is a huge opportunity to take advantage of the powerful synergistic effects of locally accessible organic materials like FYM to reduce the amount of chemical amendment (gypsum) and lower the overall cost of reclamation.

□ Crops and Cropping Sequences

In the early stages of reclamation, good crop selection is extremely important because crops differ greatly in their tolerance to soil sodicity. This topic has been the subject of extensive inquiry in India and other areas of the world. According to physiological research, preventing Na uptake and maintaining a high K: Na ratio in the shoot part aid in plant tolerance. The chosen crops should have a reclamation effect on the soil in addition to being tolerant. *Dhaincha* (*Sesbania aculeata*) green manure crop was recommended in the past as the first crop to be grown because of its tolerance to sodicity. Nevertheless, owing to the poorer establishment of *Sesbania* in highly deteriorated sodic soils, limited biomass production, lower reclaiming influence, and practically no economic returns, the recommendations for the Indo-Gangetic plain were modified subsequently.

The ideal crop for sodic soils is rice because it has a very high tolerance for it, a deep, shallow root system, requires few amendments, can easily achieve a special water regime of submergence in sodic soils with lower permeability, and can quickly make native Ca available to replace exch. Na through root activity. It also has a noticeable reclaiming effect [7, 13] Table (**3**).

Table 3. Promising tolerant crop varieties developed/identified by CSSRI, Karnal.

Crop	Varieties	Tolerance level
Rice	CSRR 10, 11	pH 9.8–10.2; ECe 10 dSm^{-1}
	CSR 12, 13, 19, 20, 21, 24, 26	pH 9.4 – 9.8
	Most varieties of normal soil	pH < 9.4; ECe < 8.5
Wheat	KRL 1-4, WH 157, Raj. 3077	pH 9.2 – 9.3; ECe 6.5
	HD2009, 2285, 2329 WH 542, C 306	pH 8.7 – 9.0 ECe 5.5

(Table 3) cont.....

Crop	Varieties	Tolerance level
Barley	CSB 1, 2, 3, Ratna DL 200, 348, BH 97	Upto pH 9.3 Upto Ece 11.0
Indian mustard	Pusa Bold, Varuna, Kranti	pH 8.8 – 9.2; ECe 6.5
Sugarbeet	Ramonaskaya – 06, Polyrava – E Tribal, Maribo-Resistapoly	pH 9.5 - 10.0 ECe 10
Sugarcane	CO 453, 1341, 6801, 62329, 1111	PH < 9.0

☐ Water Management

Drainage of sodic lands with high ESP presents special problems. Due to extremely poor water transmission characteristics, the tile subsurface drainage was found ineffective to control the water table and removing the salts at Karnal. Vertical drainage through the large-scale installation of tube wells in the sodic soils having good quality groundwater at 10-20 m depth has been effective in controlling water tables and also in augmenting irrigation resources. Based on an analysis of maximum rainfall storms as well as dry spells for Karnal conditions, Narayana (1979) [15] recommended a three-tier system of rainwater management: (i) collecting permissible water in the cropland, *e.g.*, 15 cm rainfall could be stored in the rice field, (ii) after collection of rainwater in the cropland, excess water could be stored in a dugout pond in a nearby low-lying area, and the water stored could be used for irrigation in the adjacent cropland during the dry spells by pumping, and (iii) remaining excess rainwater could be discharged into the regional surface shallow drains. Thus, the storage of rainwater in the dugout ponds serves the dual purpose of drainage and water conservation for irrigation.

☐ Phytoremediation

Phytoremediation technologies use living plants to clean up contaminated soil, air, and water. It is defined as "the use of green plants and associated microorganisms, along with proper soil amendments and agronomic techniques to either contain, remove, or render toxic environmental contaminants harmless Fig (**4**)."

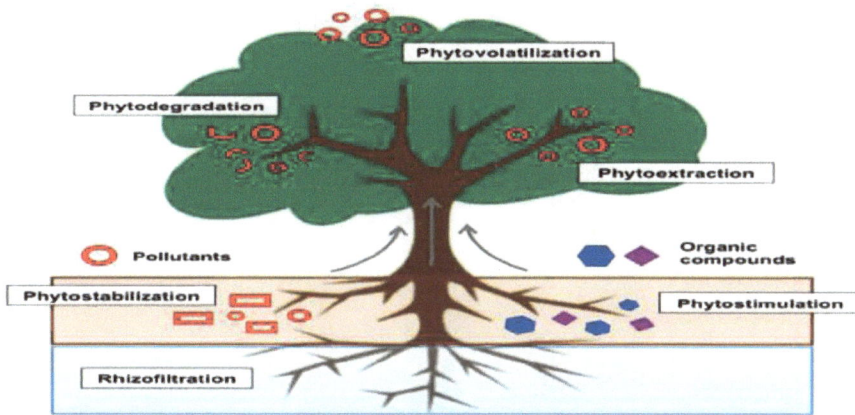

Fig. (4). Phytoremediation for managing salt affected soil.

☐ Agroforestry System

Agroforestry systems comprising a combination of *P. juliflora* and *Leptochloa fusca* brought a greater reduction in soil sodicity and a higher increase in organic C, nitrogen, infiltration rate, biological activity, and fertility status as compared to the sole plantation of *P. juliflora* [16]. Guava (*Psidium guajava*) and ber (*Zizyphus mauritiana*) models among the fruit crops and guava + Eucalyptus + subabool (*Leucena leucocephala*) among the fruit and forestry models were found more effective in improving the properties of sodic soil after five-year growth at Faizabad [17]. The corresponding reduction in pH, EC, and ESP, and increase in organic C content and hydraulic conductivity (H.C.) were noted. The ameliorative effect decreased with depth.

☐ Alternate Land Use System

In salt-affected soils, the productivity of conventional crops and cropping methods never reaches the maximum level of production. Several variables, including salinity stress, hinder agricultural productivity in severely salt-affected locations. Under these circumstances, researchers found that forestry and agroforestry are superior solutions, helping to improve the biological quality of marginal lands. Reclamation through tree plantation is a slower process, but substantial improvement in the soil is discernible in the long run. Yadav and Singh (1986) [18] observed an increase in organic carbon and nitrogen content, and a decrease in pH value and salt concentration in the upper soil under *Prosopis juliflora* plantation. Yadav and Singh (1986) [18] found *Casuarina equisetifolia* fairly tolerant to both soil sodicity and salinity. In a field experiment on sodic soil (pH 10.2-10.5, ECe 0.45-1.75 dS/m) started in 1970 [18], the ameliorative effect of different tree species was evaluated after 20 years of growth (Singh and Gill, 1992). Significant reduction in pH and EC values, and an increase in organic carbon and available P and K were noticed, though the magnitude of soil improvement varied with the species depending upon the quantity, quality, and decomposition rate of litterfall, nutrient recycling, rooting pattern and growth behavior. The amount of organic carbon and available P and K was maximum under *P. juliflora*. On the whole, the improvement was greater in the upper soil layer than in the lower layers.

5.3. Management of Saline Soil

☐ Salt Leaching and Scrapping:

Salt leaching refers to the leaching of the salts along with water, whereas salt scraping refers to a technique that includes scrubbing away the soil's top layer after the salt has built up there due to evaporation.

☐ Water Management:

For the reclamation of saline soils, adequate drainage to remove the soluble salts and to lower the water table below the critical depth, along with a proper selection of crops and varieties, balanced nutrient management, and other agronomic practices, are crucial. Both vertical and horizontal drainage systems are in use primarily to lower the water table. However, in saline soils with high saline groundwater, vertical drainage has not been favored. Therefore, horizontal subsurface drainage has received greater scientific attention. On the whole, the tiles are placed at relatively closer spacing and shallower depth to be more effective in the heavy-textured soils as compared to the light-textured soils.

☐ Bio-saline Agriculture:

The cultivation and growth of plants in saline-rich soil or groundwater are known as biosaline agriculture. Due to its toxicity to the majority of plants, salinity poses a serious threat to agriculture in areas with limited water supplies. If we consider halophytes as species that can thrive and tolerate salinity concentrations of 80 mM [19], the global halophyte plant community appears composed of more than 2600 species distributed on all the world's continents except Antarctica [19]. These comprise halophytes present in several types of ecosystems, ranging from salt deserts to mangroves, coastal dunes, and salt marshes. The success of these plants in colonizing such different environments, from arid to flooded landscapes, results from their polyphyletic origin, and thus, is a recent mechanism, such as the appearance of other complex traits such as the C4 and crassulacean acid metabolism photosynthesis pathways; this origin resulted in multiple and independent events where salt tolerance evolved in different angiosperm families. This resulted in a multitude of physiological mechanisms, such as ion compartmentation, K/Na selectivity, development of salt glands, and production of osmocompatible solutes, among others, ultimately resulting in salt-tolerance traits. Moreover, the ability of these plants to colonize both flooded and arid environments results from the same physiological tolerance basis Fig (**5**).

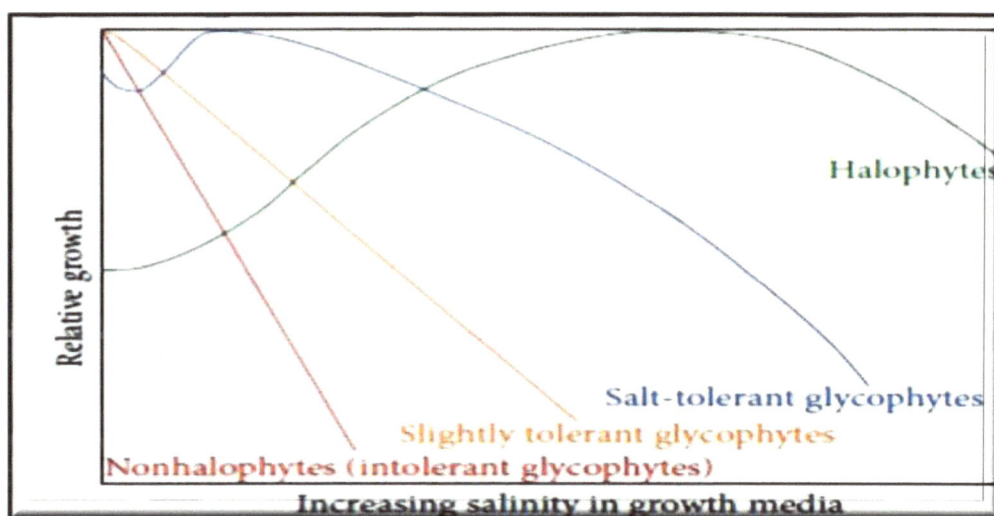

Fig. (5). Schematic illustration of growth of different kinds of plants under saline condition. [20]

CONCLUSION

- Organic residues, manures, paper mill sludge, biogas slurry, bio-char, *etc.*, help to reduce soil acidity and make the soil more productive.
- Drainage and leaching of salts from the root zone depth are important for salt-affected soils.
- The use of green manure, and different chemical and organic amendments are also helpful in managing salt-affected soils.
- New approaches like bio-saline agriculture, phytoremediation, *etc.*, can be an important strategy for reclaiming pH-related problem soils.

PATH AHEAD

- Need new and effective agronomic intervention for restoration of acid soil.
- Development of new tolerant varieties.

- Require more attention to integrated management practices.
- More research on the micro-biological consortium can improve problem soil dynamics.
- Need a proper extension to popularize new techniques to the farmers.

REFERENCES

[1]   Martinez-Beltran J, Manzur CL. Proceedings of the International Salinity Forum.

[2]   Mandal SC. Introduction and historical overview.Acid Soils of India. New Delhi, India: ICAR 1997; pp. 3-24.

[3]   Prochnow LI. Soil acidity evaluation & management. International Plant Nutrition Institute 2013.

[4]   Chintala R, Mollinedo J, Schumacher TE, *et al.* Nitrate sorption and desorption in biochars from fast pyrolysis. Microporous Mesoporous Mater 2013; 179: 250-7.
       [http://dx.doi.org/10.1016/j.micromeso.2013.05.023]

[5]   Rao DLN, Ghai SK. Urease and dehydrogenase activity of alkali and reclaimed soils. Soil Res 1985; 23(4): 661-5.
       [http://dx.doi.org/10.1071/SR9850661]

[6]   6. FAO, 1993. Forest resources assessment 1990. Tropical countries. FAO Forestry Paper 112. United Nations Food and Agriculture Organization, Rome. 61 pp. + appendices and tables.

[7]   Abrol IP, Chhabra R, Gupta RK. Fresh look at the diagnostic criteria for sodic soils. InInternational Symposium on Salt Affected Soils: 18 to 21 February 1980 1980. Karnal, India: Central Soil Salinity Research Institute,[1980?].

[8]   Sharma DK, Chaudhari SK. Agronomic research in salt affected soils of India: an overview. Indian J Agron 2012; 57(3s): 175-85.

[9]   Sarkar B, Naidu R. Nutrient and water use efficiency in soil: the influence of geological mineral amendments InNutrient use efficiency: from basics to advances. New Delhi: Springer 2015; pp. 29-44.
       [http://dx.doi.org/10.1007/978-81-322-2169-2_3]

[10]   Richards LA. Diagnosis and improvement of saline and alkali soils. LWW; 1954 Aug 1.

[11]   Agarwal RR, Yadav JSP. Diagnostic techniques for the saline and alkali soils of the Indian Gangetic alluvium in Uttar Pradesh. J Soil Sci 1956; 7(1): 109-21.
       [http://dx.doi.org/10.1111/j.1365-2389.1956.tb00867.x]

[12]   Wirojanagud W, Tantemsapya N, Tantriratna P. Precipitation of heavy metals by lime mud waste of pulp and paper mill. Songklanakarin J Sci Technol 2004; 26(1): 45-53.

[13]   Yadav JS, Agarwal RR. Dynamics of soil changes in the reclamation of saline alkali soils of the Indo-Gangetic alluvium. J Indian Soc Soil Sci 1959; 7: 213-22.

[14]   Kanwar JS, Bhumbla DR, Singh NT. Studies on the reclamation of saline and sodic soils in the Punjab. Indian J Agric Sci 1965; 35: 43-51.

[15]   Narayana VVD. Rain water management for low land rice cultivation in India. J Irrig Drain Div 1979; 105(1): 87-98.
       [http://dx.doi.org/10.1061/JRCEA4.0001245]

[16]   Singh G, Singh NT, Abrol IP. Agroforestry techniques for the rehabilitation of degraded salt-affected lands in India. Land Degrad Dev 1994; 5(3): 223-42.
       [http://dx.doi.org/10.1002/ldr.3400050306]

[17]   Khanna SS. Management of Sodic Soils through Tree Plantation (A successful case study). J Indian Soc Soil Sci 1994; 42(3): 498-508.

[18]   Yadav JS, Singh K. Response of Casuarina equisetifolia to soil salinity and sodicity. J Indian Soc

Coast Agric Res 1986; 4: 1-8.

[19]   Flowers TJ, Galal HK, Bromham L. Evolution of halophytes: multiple origins of salt tolerance in land plants. Funct Plant Biol 2010; 37(7): 604-12.
[http://dx.doi.org/10.1071/FP09269]

[20]   Hasanuzzaman M, Nahar K, Alam M, *et al.* >Potential use of halophytes to remediate saline soils. 2014.

CHAPTER 10

Impact of Heat Coupled with Drought Stress on Plants

Battana Swapna[1,*], **Srinivasan Kameswaran**[1], **Mandala Ramakrishna**[1] and **Thummala Chandrasekhar**[2]

[1] *Department of Botany, Vikrama Simhapuri University College, Kavali-524201, SPSR Nellore District, Andhra Pradesh, India*

[2] *Department of Environmental Science, Yogi Vemana University, Kadapa-516005, Andhra Pradesh, India*

Abstract: Various stages of plant growth and development could greatly be affected by abiotic stresses. Among them, two significant abiotic stressors, including drought and heat, hinder crops' vegetative or reproductive growth stages, which in turn affect sustainable agriculture worldwide. The incidence of drought coupled with heat stress is increasing mainly due to global climate change. It was proved that the effect of drought coupled with heat stress is additive when compared to individual stresses. This chapter focuses on the influence of common dual-stress heat coupled with drought stress on plants. A critical understanding of how different plants respond to heat coupled with drought stress would pave the way to developing suitable agronomic management practices for better crop genotypes with improved productivity.

Keywords: Drought, Drought coupled with heat, Heat stress, Plant stress.

1. INTRODUCTION

Naturally, the climate tends to vary over time depending on the region. Intergovernmental Panel on climate change (IPCC), in its sixth climate change Assessment Report (AR6) stated that human-induced rapid climate change, apart from natural climate variability, imposes a risk to the ecosystem [1]. Climate change aggravates the occurrence, duration, severity and impacts of some types of weather events, such as increased floods, drought, long-duration heat, or dry spells. These intense extreme events have an adverse impact on agriculture and, thus, food availability [2, 3].

* **Corresponding author Battana Swapna:** Department of Botany, Vikrama Simhapuri University College, Kavali-524201, SPSR Nellore District, Andhra Pradesh, India; E-mail: swapnaivsr@gmail.com.

Jen-Tsung Chen (Ed.)

Plants are exposed to either single stresses or multiple stresses in nature. Plant molecular responses to individual or single abiotic stresses, including drought, salt stress, flooding, extreme temperatures, light, and radiation, as well as biotic interactions like pathogen attacks and herbivory, have been studied intensively [4 - 7]. In field conditions, plants may be exposed to more than one stress factor, like simultaneous salt stress and drought, UV and drought, drought and heat, and ozone and drought [8 - 12]. Pollution leads to heavy metal contamination of sites and increased carbon dioxide concentration. Some plants are exposed to elevated CO_2 and drought, elevated CO_2 and heat stress conditions [13, 14]. Plants in contaminated sites are exposed to drought and heavy metal stress, salinity and heavy metal stress [15 - 20]. Nevertheless, more attention has been paid recently to finding the molecular mechanisms beneath interactive impacts between a combination of stresses or multiple stresses which occur simultaneously or in a sequence one after the other [21]. Further, plants may be exposed to concurrent biotic and abiotic stress factors like pathogens and drought [22, 23]. The present situation of environmental factors, besides their adverse effects on all metabolic activities of plants, made several researchers concentrate their investigation on plant stress. Progress has been made in understanding the effect of the combination of stresses to some extent on plant growth. Often, plant responses to combined stresses are not similar to individual stress responses. A combination of stresses can have synergistic or antagonistic effects on plants.

Plants ought to continuously endure a set of abiotic and biotic stressors. The sensitivity of the plants to stress depends upon the intensity and temporal extent of stress in addition to the progressive stage of the plant [24]. Plants defend themselves from stressors by responding rapidly and efficiently. Their ability to cope rests on the quick perception of changes in the environment, signal transduction within the plant, and the expression of various responses for adaptation. Adaptations of plants to stress could progress agricultural production under any adverse conditions [25, 26]. Understanding the elicitation of different signaling pathways to the specific combination of stresses is important as a response to a different combination of stressors is unique and sometimes specific to species, genotypes, and cultivars.

Food crops in arid and semi-arid regions are frequently exposed to drought coupled with heat stress due to erratic precipitation and high temperatures. The co-occurrence of prolonged heat waves with severe dry spells has increased worldwide in recent decades [27 - 34], and further, climatic models predicted drought and heat wave incidence to increase in the future [35]. Simultaneous exposure to acute or prolonged drought and heat waves causes enormous losses in agricultural yield [36, 37]. So far, our current knowledge on plant responses to drought coupled with heat stress is still limited. Understanding vital plant mechanisms responding to drought coupled with heat stress is vital for the production of dual stress-tolerant crops. In this chapter, we focus on the drought coupled with heat stress associated with physiological, morphological, and molecular changes in plants.

1.1. Morpho-physiological Responses to Drought Coupled with Heat Stress

Drought concurrent with heat stress might happen either in the vegetative or reproductive stages of plants with varied effects on the yield along with biomass of different crops [37, 38]. This stress combination has a profound effect on germination, root, tiller, stem and leaf development, root architecture, photosynthesis, carbon assimilation, and dry matter production at the vegetative growth of plants. Floral initiation, pollen fertility, fertilization, panicle exertion, seed growth, and seed filling are the reproduction stages affected by heat coupled with drought stress [39, 40]. Despite that, occasionally, the adverse effects at one stage could be indemnified by the retrieval and also further development of another organ. For example, lesser root emergence was frequently indemnified with enhanced tillering. Plenty of seeds were inhibited by partly filled seeds. Low grain yield has been indemnified by improved grain quality [41]. The reactions of plants to drought coupled with heat stress are given in Fig (1).

1.2. Plant Growth

The foremost effect of stress is on the germination and seedling phase of plants. Seeds initiate biochemical changes as soon as the imbibition of water is completed. Seed imbibition of water has relied on soil water accessibility. Drought hinders imbibition and consequently leads to diminished germination rate and germination percentage. Heat stress caused by temperatures above optimum temperatures reduces total germination percentage. Drought, coupled with heat stress, even decreased the germination percentage [42]. Studies on *Arabidopsis thaliana* have shown that drought coupled with heat stress reduces plant growth more than individual stress [43, 44]. The heat coupled with drought stress, decreased the morphological parameters in maize [45]. Leaf temperature increased in combined stress conditions in maize [46]. The desert plant, *Artmesia sieberi alba,* showed a decline in vegetative growth [47].

1.3. Root System

Roots are considered important for plant yield. They provide water and nutrient uptake, anchorage to the plant, and act as storage organs. Root architecture refers to root length, spread, number, and lateral root length that help in the acquisition of resources required for root growth. Root architecture plays an important role in plants' adaptation to abiotic stress. Roots sense the stresses first as they grow underground and alter their genetic program to survive the stress [48]. Root plasticity is triggered through differences in the growth of root constituents, number, placement and extension [49]. These variations in root architecture subsequently influence shoot growth and development [50] by altering carbon allocation to shoots. Drought coupled with heat stress in maize primary roots improved the root mass ratio, root length ratio, dry weight, tissue density and reduction of branching density. Seminal lateral roots of maize are affected by reducing the dry weight, length and surface area and improving the fineness [51, 52].

1.4. Photosynthesis

A sharp decrease in chlorophyll pigment was observed in heat coupled with drought stress of *Solanum lycopersicum* [53]. A reduction in pigment composition was described in wheat exposed to this concurrent stress with a considerable rise in the extent of de-epoxidation of xanthophyll-cycle components and chlorophyll a/b ratio [54]. A higher reduction in chlorophyll content was expressed in drought coupled with heat-stressed plants in contrast to the separate stresses applied alone for *Agrostis stolonifera* [55], chickpea [40], and soybean [56]. The decrease in leaf chlorophyll concentration may have diminished photosynthesis and Rubisco activity under stress conditions.

Generally, plants living in abiotic stress conditions exhibit depressed photosynthetic activity through the disruption of Rubisco along with photosystem II impairment [57]. But, depending on plant species, drought stress, heat stress and heat coupled with drought stress may have different impacts on photosynthesis. Photosynthesis in soybean plants under heat coupled with drought stress has been impeded more than water deficit stress alone [38]. As stated by Wang *et al.* [58], the effect of heat coupled with drought stress on photosynthetic rates was very much more detrimental than the effects of individual stresses for wheat. *Nicotiana tabacum* receiving drought along with heat exhibit reduced photosynthetic rate, closure of stomata, increased respiration, and increased leaf temperature [59]. *Populus yunnanensis* exposed to this stress combination displayed a greater decrease in photosynthetic activity and an increase in ROS generation [60]. Drastic diminution in PSII efficiency was observed in grasses from Poaceae [61]. Heat stress coupled with the drought effect on *Lotus japonicus* plants revealed a similarly remarkable disturbance in PSII function [62]. Previous reports on the influence of this stress combination specified that inhibition of PSII has been more susceptible to high-temperature stress than to water deficit stress [63]. On the other hand, some studies found that heat stress has a minimum impact on drought stress in tomatoes and olives about PSII function [64, 65]. These contrasting results might be due to the duration of heat exposure.

Elevated rubisco activity was noticed due to high-temperature stress, while it declined regarding drought or heat coupled with drought stress in *Cicer arietinum* [40]. Two cultivars of cotton with different sensitivity to drought showed severe reduction in photosynthetic activity and stomatal conductance when exposed to individual high-temperature stress or heat coupled with drought stress. This resulted in a more pronounced effect during the combined stress conditions. This stress combination imposed lower Rubisco activity in drought-sensitive cultivars when compared to drought-tolerant cultivars [66]. According to Abdelhakim *et al.* [54], heat stress-tolerant and heat-sensitive genotypes of wheat presented different sensitivities to the single and concurrent heat stress and drought during photosynthesis. During early combined stress, Fq'/Fm' has not been drastically affected as the effect of drought was not severe [67]. With the sharp rise in stress severity and duration, the electron transport rate decreased and, subsequently, the capacity of the photosynthetic apparatus to sustain oxidized QA. In the later combined stress phase, Fq'/Fm' declined rapidly since the effect of drought was pronounced. However, the heat-sensitive genotype showed higher Fq'/Fm' when compared to heat-tolerant genotypes. During recovery, the heat-tolerant genotype performed better than the heat-sensitive genotype, showing that the damage caused by stress was not severe. Exposure of citrus to heat coupled with drought stress demonstrated that the impact of water deficit stress on gas exchange parameters dominated over high-temperature stress. It was confirmed through the induction of stomatal closure prevailing over changes that led to the decreased surface temperature of the leaf [68]. Canola plants exhibited a conservation strategy response through stomata closing rather than leaf cooling. Under heat stress coupled with drought conditions, they expressed low g_m compared to drought stress [69]. Findings from the above study propose that photosynthesis plays a significant part in plant adaptation to heat stress coupled with drought Fig (**1**). Both drought- and heat-induced restrictions turn at the same time to constrain the photosynthetic activity of plants exposed to concurrent or simultaneous stresses in nature.

Multiple ultrastructural abnormalities of chloroplasts and mitochondria in leaf mesophyll cells happened depending on the varietal difference, as reported by Grigorova *et al.* [70].

A combination of stress in soybean plants displayed stomatal response over heat stress at both flowering and pod stages. Hence, to maintain a balance between the prevention of water loss and protection from heat stress, it seems that stomatal closure prevails over transpiration loss for cooling the leaf surface. Similar data in soybean have been demonstrated recently by Katam *et al.* [71]. Vile *et al.* [44] proved that *Arabidopsis* plants exposed to water deficit stress improved stomatal density while heat showed an antagonistic effect. Furthermore, stomatal density in plants undergoing combined stresses diminished, signifying that intense stress determined the plant's metabolic activities.

1.5. Metabolites

The metabolite profile of plants exposed to heat stress coupled with drought has expressed several metabolites comprising compatible osmolytes, sugars, polyols, and intermediates of the Krebs cycle specifically altered [23]. Proline has accumulated in drought-stressed *Arabidopsis thaliana* plants, whereas it is not accumulated during individual high-temperature stress or the heat stress coupled with drought exposure. On the contrary, plants accumulated sucrose when exposed to heat coupled with drought stress to defend cellular components against the negative impact of drought [43]. Many studies propose that sucrose might act as an osmoprotectant instead of proline in *Arabidopsis* plants exposed to heat coupled with drought stress. In the same manner, *Portulaca* also exhibited sucrose accumulation [72]. However, in tobacco plants, Cvikrova *et al.* [73] proved the plausible proline contribution in the defense against heat coupled with drought stress through the modification of polyamine biosynthesis.

1.6. Antioxidants

The stress effect on antioxidative metabolism is frequently related to electron leakage and initiation of oxidative damage [74]. Reactive oxygen species (ROS) comprising hydrogen peroxide (H_2O_2), superoxide anion (O_2^-), and hydroxyl radical (•OH) are generated in cells upon exposure to abiotic stresses. ROS are generated mainly in organelles like chloroplasts, peroxisomes, and mitochondria [75]. Individual drought and heat stress considerably raise ROS levels that lead to oxidative damage of macromolecules [76, 77]. Specifically, ROS attacks membrane lipids causing lipid peroxidation, and the malondialdehyde (MDA) content is enhanced. MDA was considered one of the markers of oxidative damage in plants [78]. In numerous biological processes, including stomatal closure, growth, development, and stress signaling, ROS serve as signaling molecules [79].

The antioxidative defenses are considered crucial factors for stress tolerance [80]. The heat coupled with drought stress impact on antioxidant activities of Buffel grass genotypes showed that oxidative damage was higher in the sensitive genotype. This data put forward that oxidative damage could be associated with species' vulnerability to heat coupled with drought stress [74]. Several other studies are following this result [78, 80, 81].

1.7. Yield

The heat coupled with drought stress, interrupts the inception of flowering stages in sorghum [82, 83]. The combined stress at the pre-flowering stage showed consequences like abortion of the florets, and a reduction in panicle size and size of the seed leads to decreased yield [84, 85]. Pollination and fertilization are dramatically impacted by heat coupled with drought stress during sorghum reproductive stages, instigated by reduced pollen viability and receptivity of the stigma [86 - 89]. The simultaneous drought and heat stress have a cumulative effect on seed yields that are substantially greater than single stress effects in soybean [90 - 92]. Cereals like barley [67] and wheat [90, 93] respond similarly. Heat stress coupled with drought applied during the blooming period exhibited a decrease in grain yield in tomatoes [94], groundnut [40, 95], and lentil [96]. Seed yield is chiefly dependent on pod numbers (PN), thousand seed weights (TSW), and the number of seeds (SN). Seed yield has been reduced in soybean due to PN and SN reductions. Further, variation in seed yield has been elucidated by the changes in TSW under combined stresses. Similar results have been reported by Ergo *et al*. [97].

Drought, together with heat stress, induced lesser biomass and earlier flowering when compared to single stresses. Drought and heat stress significantly increased ovule abortion [39]. The heat coupled with drought stress, negatively affects floral evocation and seed formation by impeding sucrose transportation during the emergence of the ear. Floret number in the base reduced when exposed to heat stress, whereas floret number in the top regions significantly diminished with drought stress [98]. The heat coupled with drought stress caused an alteration of gametogenesis, pistil anatomy, and a reduction in pollen viability. Nevertheless, the concurrent stress afflicted male reproductive organs in wheat [99, 100].

Schmidt *et al*. [101] assessed wheat accessions for grain development utilizing X-ray computed tomography under heat coupled with drought stress. Hollow cavities are formed in seeds with combined stress showing inadequate development as well as inhibition of grain filling. Zhang *et al*. [102] described that protein content amplified in wheat grains when combined stress was applied at the grain-filling stage. Also, a consistent rise in grain protein in heat coupled with drought stress was noticed. Thus, the effects of heat coupled with drought stress on grain proteins remain mostly elusive and need further study.

1.8. Molecular Responses to Heat Coupled with Drought Stress

Transcriptome analysis of plants exposed to heat coupled with drought stress showed a new profile of transcript expression, which was not projected by the individual stresses [103]. Various crop plants expressed unique transcriptome changes under the stress combination. Almost 770 unique transcripts that encode several heat shock proteins, protein kinases, enzymes associated with lipid biosynthesis and proteases, were expressed in *Arabidopsis thaliana* plants treated with single stresses drought, heat, and combined stress heat coupled with drought stress. But not changed with exposure to drought or heat stress alone [43].

Heat shocks protein-coding transcripts, such as small HSP70, HSPs, HSP100, and HSP90, in addition to transcripts encoding proteins related to phenylalanine ammonia lyase, pathogenesis, WRKY transcription factors, and multiprotein bridging factor (MBF1c), were upregulated in tobacco plants exposed to heat coupled with drought stress [59]. In the course of drought, antioxidative enzymes, including catalase (CAT) and glutathione peroxidase (GPX), was triggered, whereas high-temperature stress elevated ascorbate peroxidase (APX) and thioredoxin peroxidase (TPX). Plants confronted with drought stress and combination particularly produce transcripts encoding superoxide dismutase (SOD), glutathione peroxidase (GPX), glutathione S transferase (GST), glutathione reductase (GR) and alternative oxidase (AOX). Additionally, Koussevitzky *et al.* [104] observed that the ascorbate peroxidase gene was especially necessary for *Arabidopsis thaliana* plants to become tolerant to heat and drought stress. Stronger antioxidative capacity can be related to a plant's ability to resist simultaneous stresses since initiation of ROS detoxifying enzymes was considered a general response amongst plant species to heat coupled with drought stress.

Subsequently, transcription factors (TFs) like MYB [43], HSPs, and transcripts implicated in photosynthesis and respiration are concentrated in response to heat coupled with drought. Plants have to keep up equilibrium between energy and resource distribution for stress adaptation. Plant molecular responses to stressful conditions may be contemplated in the total transcripts changed by each stress condition alone or in combination. In sorghum plants, for instance, a transcriptome study of single and dual stress conditions found drought stress regulation of 448 transcripts, heat regulation of 1554 transcripts, and 2043 transcripts regulated by their combination [105], representing that a greater number of transcripts, particularly changed when drought together with heat stress is present relative to each condition present alone.

1.9. New Approaches for Developing Tolerance to Heat Coupled with Drought Stress

Certain new approaches for elucidating plant responses to heat coupled with drought stress are immediately needed to develop tolerant varieties against dual stress. These studies considerably increase our probability of creating crops that are more tolerant to environmental conditions in the field by revealing the intricate networks of molecular interactions. Developmental studies blended "omic" studies might decide the outline of gene expression and unveil the molecular basis for the consequences of heat stress coupled with drought. In the way to assume processes involved in the responses to stress, diverse methods relating to transcriptomics, micromics, genomics, epigenomics, and proteomics are necessary to recognize the crucial genes and other regulatory molecules or proteins required for genetic improvement. This can present positive leads to developing new approaches for improving tolerance.

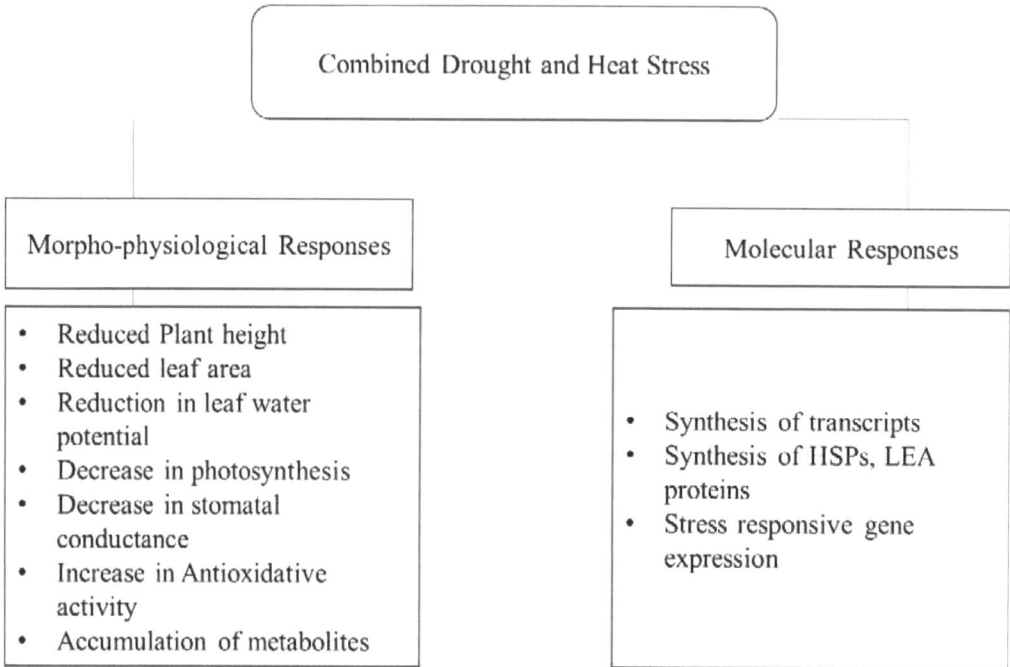

Fig. (1). Responses of plants to heat coupled with drought stress.

CONCLUSION AND FUTURE PERPSPECTIVES

The heat, coupled with drought stress, influences many developmental as well as physiological activities of crop plants. The effects depend on stress severity along with duration and also the time of exposure during the plant life cycle. Crop plants show differential responses to the combination of stresses. The effect of heat coupled with drought stress was synergistic in some species and less additive in some species. Combined stress impact is prominent during the reproductive phase.

Further, thorough studies are essential to reveal the mechanisms underlying the responses to heat coupled with drought stress. It is necessary to examine possible interactions by modeling water deficit stress and high-temperature stress impacts simultaneously. A consistent water deficit and temperature threshold causing spikelet sterility of rice in field conditions, including cultivars with extremely variable phenology, have been a major limitation for rice breeders to formulate acclimation strategies. The identification and deployment of alleles for heat coupled with drought stress tolerance in crop breeding are necessary to design new varieties that can fulfill the rapidly growing population due to climate change scenarios. The performance of transgenic plants under heat coupled with drought stress has to be tested in field circumstances. An all-inclusive approach integrating various management practices to deal with heat coupled with drought stress might be a win-win approach.

REFERENCES

[1]   Portner HO, Roberts DC, Poloczanska ES, Eds., *et al.* 2022.

[2]   Bellard C, Bertelsmeier C, Leadley P, Thuiller W, Courchamp F. Impacts of climate change on the future of biodiversity. Ecol Lett 2012; 15(4): 365-77.
[http://dx.doi.org/10.1111/j.1461-0248.2011.01736.x] [PMID: 22257223]

[3]   Rosenzweig C, Elliott J, Deryng D, *et al.* Assessing agricultural risks of climate change in the 21st century in a global gridded crop model intercomparison. Proc Natl Acad Sci USA 2014; 111(9): 3268-73.
[http://dx.doi.org/10.1073/pnas.1222463110] [PMID: 24344314]

[4]   Kaur G, Asthir B. Molecular responses to drought stress in plants. Biol Plant 2017; 61(2): 201-9.
[http://dx.doi.org/10.1007/s10535-016-0700-9]

[5]   Zhao S, Zhang Q, Liu M, Zhou H, Ma C, Wang P. Regulation of Plant Responses to Salt Stress. Int J Mol Sci 2021; 22(9): 4609.
[http://dx.doi.org/10.3390/ijms22094609] [PMID: 33924753]

[6]   Fukao T, Barrera-Figueroa BE, Juntawong P, Peña-Castro JM. Submergence and Waterlogging Stress in Plants: A Review Highlighting Research Opportunities and Understudied Aspects. Front Plant Sci 2019; 10: 340.
[http://dx.doi.org/10.3389/fpls.2019.00340] [PMID: 30967888]

[7]   Akter N, Rafiqul Islam M. Heat stress effects and management in wheat. A review. Agron Sustain Dev 2017; 37(5): 37.
[http://dx.doi.org/10.1007/s13593-017-0443-9]

[8]   Fahad S, Bajwa AA, Nazir U, *et al.* Crop Production under Drought and Heat Stress: Plant Responses and Management Options. Front Plant Sci 2017; 8: 1147.
[http://dx.doi.org/10.3389/fpls.2017.01147] [PMID: 28706531]

[9]   Dugasa MT, Cao F, Ibrahim W, Wu F. Differences in physiological and biochemical characteristics in response to single and combined drought and salinity stresses between wheat genotypes differing in salt tolerance. Physiol Plant 2019; 165(2): 134-43.
[http://dx.doi.org/10.1111/ppl.12743] [PMID: 29635753]

[10]   Paul K, Pauk J, Kondic-Spika A, *et al.* Co-occurrence of Mild Salinity and Drought Synergistically Enhances Biomass and Grain Retardation in Wheat. Front Plant Sci 2019; 10: 501.
[http://dx.doi.org/10.3389/fpls.2019.00501] [PMID: 31114595]

[11]   Jan R, Khan MA, Asaf S, *et al.* Drought and UV Radiation Stress Tolerance in Rice Is Improved by Overaccumulation of Non-Enzymatic Antioxidant Flavonoids. Antioxidants 2022; 11(5): 917.
[http://dx.doi.org/10.3390/antiox11050917] [PMID: 35624781]

[12]   Maliba BG, Inbaraj PM, Berner JM. The effect of ozone and drought on the photosynthetic performance of canola. J Integr Agric 2018; 17(5): 1137-44.

[http://dx.doi.org/10.1016/S2095-3119(17)61834-3]

[13]   Barickman TC, Adhikari B, Sehgal A, Walne CH, Reddy KR, Gao W. Drought and Elevated CO2 Impacts Photosynthesis and Biochemicals of Basil (Ocimum basilicum L.). Stresses 2021; 1(4): 223-37.
[http://dx.doi.org/10.3390/stresses1040016]

[14]   Zhou R, Yu X, Wen J, *et al.* Interactive effects of elevated CO_2 concentration and combined heat and drought stress on tomato photosynthesis. BMC Plant Biol 2020; 20(1): 260.
[http://dx.doi.org/10.1186/s12870-020-02457-6] [PMID: 32505202]

[15]   Bashir W, Anwar S, Zhao Q, Hussain I, Xie F. Interactive effect of drought and cadmium stress on soybean root morphology and gene expression. Ecotoxicol Environ Saf 2019; 175: 90-101.
[http://dx.doi.org/10.1016/j.ecoenv.2019.03.042] [PMID: 30889404]

[16]   Ellouzi H, Ben Hamed K, Asensi-Fabado MA, Müller M, Abdelly C, Munné-Bosch S. Drought and cadmium may be as effective as salinity in conferring subsequent salt stress tolerance in Cakile maritima. Planta 2013; 237(5): 1311-23.
[http://dx.doi.org/10.1007/s00425-013-1847-7] [PMID: 23381736]

[17]   de Silva NDG, Cholewa E, Ryser P. Effects of combined drought and heavy metal stresses on xylem structure and hydraulic conductivity in red maple (Acer rubrum L.). J Exp Bot 2012; 63(16): 5957-66.
[http://dx.doi.org/10.1093/jxb/ers241] [PMID: 22966005]

[18]   Gul S, Nawaz MF, Azeem M. Interactive effects of salinity and heavy metal stress on ecophysiological responses of two maize (Zea mays L.) cultivars. FUUAST Journal of Biology 2016; 6(1): 81-7.

[19]   Naz R, Sarfraz A, Anwar Z, *et al.* Combined ability of salicylic acid and spermidine to mitigate the individual and interactive effects of drought and chromium stress in maize (Zea mays L.). Plant Physiol Biochem 2021; 159: 285-300.
[http://dx.doi.org/10.1016/j.plaphy.2020.12.022] [PMID: 33418188]

[20]   Ouni Y, Mateos-Naranjo E, Abdelly C, Lakhdar A. Interactive effect of salinity and zinc stress on growth and photosynthetic responses of the perennial grass, Polypogon. Ecol Eng 2016; 95: 171-9.
[http://dx.doi.org/10.1016/j.ecoleng.2016.06.067]

[21]   Pereira A. Plant Abiotic Stress Challenges from the Changing Environment. Front Plant Sci 2016; 7: 1123.
[http://dx.doi.org/10.3389/fpls.2016.01123] [PMID: 27512403]

[22]   Rejeb I, Pastor V, Mauch-Mani B. Plant Responses to Simultaneous Biotic and Abiotic Stress: Molecular Mechanisms. Plants 2014; 3(4): 458-75.
[http://dx.doi.org/10.3390/plants3040458] [PMID: 27135514]

[23]   Suzuki N, Rivero RM, Shulaev V, Blumwald E, Mittler R. Abiotic and biotic stress combinations. New Phytol 2014; 203(1): 32-43.
[http://dx.doi.org/10.1111/nph.12797] [PMID: 24720847]

[24]   Mittler R, Blumwald E. Genetic engineering for modern agriculture: challenges and perspectives. Annu Rev Plant Biol 2010; 61(1): 443-62.
[http://dx.doi.org/10.1146/annurev-arplant-042809-112116] [PMID: 20192746]

[25]   Nguyen D, Rieu I, Mariani C, van Dam NM. How plants handle multiple stresses: hormonal interactions underlying responses to abiotic stress and insect herbivory. Plant Mol Biol 2016; 91(6): 727-40.
[http://dx.doi.org/10.1007/s11103-016-0481-8] [PMID: 27095445]

[26]   Yoshida T, Mogami J, Yamaguchi-Shinozaki K. ABA-dependent and ABA-independent signaling in response to osmotic stress in plants. Curr Opin Plant Biol 2014; 21: 133-9.
[http://dx.doi.org/10.1016/j.pbi.2014.07.009] [PMID: 25104049]

[27]   Ciais P, Reichstein M, Viovy N, *et al.* Europe-wide reduction in primary productivity caused by the heat and drought in 2003. Nature 2005; 437(7058): 529-33.

[http://dx.doi.org/10.1038/nature03972] [PMID: 16177786]

[28]   van Dijk AIJM, Beck HE, Crosbie RS, *et al.* The Millennium Drought in southeast Australia (2001-2009): Natural and human causes and implications for water resources, ecosystems, economy, and society. Water Resour Res 2013; 49(2): 1040-57.
[http://dx.doi.org/10.1002/wrcr.20123]

[29]   Mazdiyasni O, AghaKouchak A. Substantial increase in concurrent droughts and heatwaves in the United States. Proc Natl Acad Sci USA 2015; 112(37): 11484-9.
[http://dx.doi.org/10.1073/pnas.1422945112] [PMID: 26324927]

[30]   Shukla S, Safeeq M, AghaKouchak A, Guan K, Funk C. Temperature impacts on the water year 2014 drought in California. Geophys Res Lett 2015; 42(11): 4384-93.
[http://dx.doi.org/10.1002/2015GL063666]

[31]   Tricker PJ, Haefele SM, Okamoto M. The interaction of drought and nutrient stress in wheat.Water stress and crop plants: A sustainable approach. Chichester: John Wiley & Sons, Ltd 2016; pp. 695-710.
[http://dx.doi.org/10.1002/9781119054450.ch40]

[32]   Panda DK, AghaKouchak A, Ambast SK. Increasing heat waves and warm spells in India, observed from a multiaspect framework. J Geophys Res Atmos 2017; 122(7): 3837-58.
[http://dx.doi.org/10.1002/2016JD026292]

[33]   Sharma S, Mujumdar P. Increasing frequency and spatial extent of concurrent meteorological droughts and heatwaves in India. Sci Rep 2017; 7(1): 15582.
[http://dx.doi.org/10.1038/s41598-017-15896-3] [PMID: 29138468]

[34]   Kong Q, Guerreiro SB, Blenkinsop S, Li XF, Fowler HJ. Increases in summertime concurrent drought and heatwave in Eastern China. Weather Clim Extrem 2020; 28: 100242.
[http://dx.doi.org/10.1016/j.wace.2019.100242]

[35]   Zscheischler J, Westra S, van den Hurk BJJM, *et al.* Future climate risk from compound events. Nat Clim Chang 2018; 8(6): 469-77.
[http://dx.doi.org/10.1038/s41558-018-0156-3]

[36]   Chen S, Guo Y, Sirault X, *et al.* Nondestructive Phenomic Tools for the Prediction of Heat and Drought Tolerance at Anthesis in Brassica Species. Plant Phenomics 2019; pp. 1-16.

[37]   Zandalinas SI, Fritschi FB, Mittler R. Signal transduction networks during stress combination. J Exp Bot 2020; 71(5): 1734-41.
[http://dx.doi.org/10.1093/jxb/erz486] [PMID: 31665392]

[38]   Cohen I, Zandalinas SI, Huck C, Fritschi FB, Mittler R. Meta-analysis of drought and heat stress combination impact on crop yield and yield components. Physiol Plant 2020; 487.
[PMID: 32880977]

[39]   Pradhan GP, Prasad PVV, Fritz AK, Kirkham MB, Gill BS. Effects of drought and high temperature stress on synthetic hexaploid wheat. Funct Plant Biol 2012; 39(3): 190-8.
[http://dx.doi.org/10.1071/FP11245] [PMID: 32480773]

[40]   Awasthi R, Kaushal N, Vadez V, *et al.* Individual and combined effects of transient drought and heat stress on carbon assimilation and seed filling in chickpea. Funct Plant Biol 2014; 41(11): 1148-67.
[http://dx.doi.org/10.1071/FP13340] [PMID: 32481065]

[41]   Prasad PV, Staggenborg SA, Ristic Z. Impacts of drought and/or heat stress on physiological, developmental, growth, and yield processes of crop plants. Response of crops to limited water: Understanding and modeling water stress effects on plant growth processes. 2008; 1: 301-55.

[42]   Blanchard-Gros R, Bigot S, Martinez JP, Lutts S, Guerriero G, Quinet M. Comparison of drought and heat resistance strategies among six populations of *Solanum chilense* and two cultivars of *Solanum lycopersicum.* Plants 2021; 10(8): 1720.
[http://dx.doi.org/10.3390/plants10081720] [PMID: 34451764]

[43] Rizhsky L, Liang H, Shuman J, Shulaev V, Davletova S, Mittler R. When defense pathways collide. The response of Arabidopsis to a combination of drought and heat stress. Plant Physiol 2004; 134(4): 1683-96.
[http://dx.doi.org/10.1104/pp.103.033431] [PMID: 15047901]

[44] Vile D, Pervent M, Belluau M, *et al.* Arabidopsis growth under prolonged high temperature and water deficit: independent or interactive effects? Plant Cell Environ 2012; 35(4): 702-18.
[http://dx.doi.org/10.1111/j.1365-3040.2011.02445.x] [PMID: 21988660]

[45] Hussain HA, Men S, Hussain S, *et al.* Interactive effects of drought and heat stresses on morpho-physiological attributes, yield, nutrient uptake and oxidative status in maize hybrids. Sci Rep 2019; 9(1): 3890.
[http://dx.doi.org/10.1038/s41598-019-40362-7] [PMID: 30846745]

[46] Tandzi LN, Bradley G, Mutengwa C. Morphological responses of maize to drought, heat and combined stresses at seedling stage. J Biol Sci (Faisalabad, Pak) 2018; 19(1): 7-16.
[http://dx.doi.org/10.3923/jbs.2019.7.16]

[47] Alhaithloul HAS. Impact of Combined Heat and Drought Stress on the Potential Growth Responses of the Desert Grass *Artemisia sieberi alba*: Relation to Biochemical and Molecular Adaptation. Plants 2019; 8(10): 416.
[http://dx.doi.org/10.3390/plants8100416] [PMID: 31618849]

[48] Lynch J. Root architecture and plant productivity. Plant Physiol 1995; 109(1): 7-13.
[http://dx.doi.org/10.1104/pp.109.1.7] [PMID: 12228579]

[49] Giehl RFH, Gruber BD, von Wirén N. It's time to make changes: modulation of root system architecture by nutrient signals. J Exp Bot 2014; 65(3): 769-78.
[http://dx.doi.org/10.1093/jxb/ert421] [PMID: 24353245]

[50] Paez-Garcia A, Motes C, Scheible WR, Chen R, Blancaflor E, Monteros M. Root traits and phenotyping strategies for plant improvement. Plants 2015; 4(2): 334-55.
[http://dx.doi.org/10.3390/plants4020334] [PMID: 27135332]

[51] Chimungu JG, Loades KW, Lynch JP. Root anatomical phenes predict root penetration ability and biomechanical properties in maize (Zea Mays). J Exp Bot 2015; 66(11): 3151-62.
[http://dx.doi.org/10.1093/jxb/erv121] [PMID: 25903914]

[52] Vescio R, Abenavoli MR, Sorgonà A. Single and combined abiotic stress in maize root morphology. Plants 2020; 10(1): 5.
[http://dx.doi.org/10.3390/plants10010005] [PMID: 33374570]

[53] Raja V, Qadir SU, Alyemeni MN, Ahmad P. Impact of drought and heat stress individually and in combination on physio-biochemical parameters, antioxidant responses, and gene expression in *Solanum lycopersicum.* 3 Biotech 2020; 10(5): 208.
[http://dx.doi.org/10.1007/s13205-020-02206-4] [PMID: 32351866]

[54] Abdelhakim LOA, Rosenqvist E, Wollenweber B, Spyroglou I, Ottosen CO, Panzarová K. Investigating combined drought-and heat stress effects in wheat under controlled conditions by dynamic image-based phenotyping. Agronomy (Basel) 2021; 11(2): 364.
[http://dx.doi.org/10.3390/agronomy11020364]

[55] McCann SE, Huang B. Effects of trinexapac-ethyl foliar application on creeping bentgrass responses to combined drought and heat stress. Crop Sci 2007; 47(5): 2121-8.
[http://dx.doi.org/10.2135/cropsci2006.09.0614]

[56] Jumrani K, Bhatia VS. Interactive effect of temperature and water stress on physiological and biochemical processes in soybean. Physiol Mol Biol Plants 2019; 25(3): 667-81.
[http://dx.doi.org/10.1007/s12298-019-00657-5] [PMID: 31168231]

[57] Nishiyama Y, Murata N. Revised scheme for the mechanism of photoinhibition and its application to enhance the abiotic stress tolerance of the photosynthetic machinery. Appl Microbiol Biotechnol 2014;

98(21): 8777-96.
[http://dx.doi.org/10.1007/s00253-014-6020-0] [PMID: 25139449]

[58]   Wang GP, Hui Z, Li F, Zhao MR, Zhang J, Wang W. Improvement of heat and drought photosynthetic tolerance in wheat by overaccumulation of glycinebetaine. Plant Biotechnol Rep 2010; 4(3): 213-22.
[http://dx.doi.org/10.1007/s11816-010-0139-y]

[59]   Rizhsky L, Liang H, Mittler R. The combined effect of drought stress and heat shock on gene expression in tobacco. Plant Physiol 2002; 130(3): 1143-51.
[http://dx.doi.org/10.1104/pp.006858] [PMID: 12427981]

[60]   Li X, Yang Y, Sun X, *et al.* Comparative physiological and proteomic analyses of poplar (*Populus yunnanensis*) plantlets exposed to high temperature and drought. PLoS One 2014; 9(9): e107605.
[http://dx.doi.org/10.1371/journal.pone.0107605] [PMID: 25225913]

[61]   Jiang Y, Huang B. Drought and heat stress injury to two cool-season turfgrasses in relation to antioxidant metabolism and lipid peroxidation. Crop Sci 2001; 41(2): 436-42.
[http://dx.doi.org/10.2135/cropsci2001.412436x]

[62]   Sainz M, Díaz P, Monza J, Borsani O. Heat stress results in loss of chloroplast Cu/Zn superoxide dismutase and increased damage to Photosystem II in combined drought-heat stressed *Lotus japonicus*. Physiol Plant 2010; 140(1): 46-56.
[http://dx.doi.org/10.1111/j.1399-3054.2010.01383.x] [PMID: 20487374]

[63]   Killi D, Raschi A, Bussotti F. Lipid peroxidation and chlorophyll fluorescence of photosystem II performance during drought and heat stress is associated with the antioxidant capacities of C3 sunflower and C4 maize varieties. Int J Mol Sci 2020; 21(14): 4846.
[http://dx.doi.org/10.3390/ijms21144846] [PMID: 32659889]

[64]   Zhou R, Kong L, Wu Z, *et al.* Physiological response of tomatoes at drought, heat and their combination followed by recovery. Physiol Plant 2019; 165(2): 144-54.
[http://dx.doi.org/10.1111/ppl.12764] [PMID: 29774556]

[65]   Haworth M, Marino G, Brunetti C, Killi D, De Carlo A, Centritto M. The impact of heat stress and water deficit on the photosynthetic and stomatal physiology of olive (*Olea europaea* L.)—A case study of the 2017 heat wave. Plants 2018; 7(4): 76.
[http://dx.doi.org/10.3390/plants7040076] [PMID: 30241389]

[66]   Carmo-Silva AE, Gore MA, Andrade-Sanchez P, French AN, Hunsaker DJ, Salvucci ME. Decreased CO_2 availability and inactivation of Rubisco limit photosynthesis in cotton plants under heat and drought stress in the field. Environ Exp Bot 2012; 83: 1-11.
[http://dx.doi.org/10.1016/j.envexpbot.2012.04.001]

[67]   Rollins JA, Habte E, Templer SE, Colby T, Schmidt J, von Korff M. Leaf proteome alterations in the context of physiological and morphological responses to drought and heat stress in barley (Hordeum vulgare L.). J Exp Bot 2013; 64(11): 3201-12.
[http://dx.doi.org/10.1093/jxb/ert158] [PMID: 23918963]

[68]   Zandalinas S. Plant strategies to deal with a combination of drought and high temperatures (Doctoral dissertation, Universitat Jaume I) 2016.

[69]   Elferjani R, Soolanayakanahally R. Canola responses to drought, heat, and combined stress: shared and specific effects on carbon assimilation, seed yield, and oil composition. Front Plant Sci 2018; 9: 1224.
[http://dx.doi.org/10.3389/fpls.2018.01224] [PMID: 30214451]

[70]   Grigorova B, Vassileva V, Klimchuk D, Vaseva I, Demirevska K, Feller U. Drought, high temperature, and their combination affect ultrastructure of chloroplasts and mitochondria in wheat (*Triticum aestivum* L.) leaves. J Plant Interact 2012; 7(3): 204-13.
[http://dx.doi.org/10.1080/17429145.2011.654134]

[71]   Katam R, Shokri S, Murthy N, *et al.* Proteomics, physiological, and biochemical analysis of cross

tolerance mechanisms in response to heat and water stresses in soybean. PLoS One 2020; 15(6): e0233905.
[http://dx.doi.org/10.1371/journal.pone.0233905] [PMID: 32502194]

[72] Jin R, Wang Y, Liu R, Gou J, Chan Z. Physiological and metabolic changes of purslane (Portulaca oleracea L.) in response to drought, heat, and combined stresses. Frontiers in Plant Science 7(6): 11-23.2016;

[73] Cvikrová M, Gemperlová L, Martincová O, Vanková R. Effect of drought and combined drought and heat stress on polyamine metabolism in proline-over-producing tobacco plants. Plant Physiol Biochem 2013; 73(73): 7-15.
[http://dx.doi.org/10.1016/j.plaphy.2013.08.005] [PMID: 24029075]

[74] Tommasino E, López Colomba E, Carrizo M, *et al.* Individual and combined effects of drought and heat on antioxidant parameters and growth performance in Buffel grass (*Cenchrus ciliaris* L.) genotypes. S Afr J Bot 2018; 119: 104-11.
[http://dx.doi.org/10.1016/j.sajb.2018.08.026]

[75] Apel K, Hirt H. Reactive oxygen species: metabolism, oxidative stress, and signal transduction. Annu Rev Plant Biol 2004; 55(1): 373-99.
[http://dx.doi.org/10.1146/annurev.arplant.55.031903.141701] [PMID: 15377225]

[76] Farooq M, Wahid A, Kobayashi NS, Fujita DB, Basra SM. Plant drought stress: effects, mechanisms and management.Sustainable agriculture. Dordrecht: Springer 2009; pp. 153-88.
[http://dx.doi.org/10.1007/978-90-481-2666-8_12]

[77] Gill SS, Tuteja N. Reactive oxygen species and antioxidant machinery in abiotic stress tolerance in crop plants. Plant Physiol Biochem 2010; 48(12): 909-30.
[http://dx.doi.org/10.1016/j.plaphy.2010.08.016] [PMID: 20870416]

[78] Bi A, Fan J, Hu Z, *et al.* Differential acclimation of enzymatic antioxidant metabolism and photosystem II photochemistry in tall fescue under drought and heat and the combined stresses. Front Plant Sci 2016; 7: 453.
[http://dx.doi.org/10.3389/fpls.2016.00453] [PMID: 27148288]

[79] Suzuki N, Koussevitzky S, Mittler R, Miller G. ROS and redox signalling in the response of plants to abiotic stress. Plant Cell Environ 2012; 35(2): 259-70.
[http://dx.doi.org/10.1111/j.1365-3040.2011.02336.x] [PMID: 21486305]

[80] Sekmen AH, Ozgur R, Uzilday B, Turkan I. Reactive oxygen species scavenging capacities of cotton (*Gossypium hirsutum*) cultivars under combined drought and heat induced oxidative stress. Environ Exp Bot 2014; 99: 141-9.
[http://dx.doi.org/10.1016/j.envexpbot.2013.11.010]

[81] Zandalinas SI, Balfagón D, Arbona V, Gómez-Cadenas A. Modulation of antioxidant defense system is associated with combined drought and heat stress tolerance in citrus. Front Plant Sci 2017; 8: 953.
[http://dx.doi.org/10.3389/fpls.2017.00953] [PMID: 28638395]

[82] Wenzel WG. Effect of moisture stress on sorghum yield and its components. S Afr J Plant Soil 1999; 16(3): 153-7.
[http://dx.doi.org/10.1080/02571862.1999.10635002]

[83] Munamava M, Riddoch I. Response of three sorghum (*Sorghum bicolor* L. Moench) varieties to soil moisture stress at different developmental stages. S Afr J Plant Soil 2001; 18(2): 75-9.
[http://dx.doi.org/10.1080/02571862.2001.10634407]

[84] Jabereldar AA, El Naim AM, Abdalla AA, Dagash YM. Effect of water stress on yield and water use efficiency of sorghum (*Sorghum bicolor* L. Moench) in semi-arid environment. Int J Agric For 2017; 7(1): 1-6.

[85] Ndlovu E, van Staden J, Maphosa M. Morpho-physiological effects of moisture, heat and combined stresses on *Sorghum bicolor* [Moench (L.)] and its acclimation mechanisms. Plant Stress 2021; 2:

100018.
[http://dx.doi.org/10.1016/j.stress.2021.100018]

[86]   Barnabás B, Jäger K, Fehér A. The effect of drought and heat stress on reproductive processes in cereals. Plant Cell Environ 2008; 31(1): 11-38.
[PMID: 17971069]

[87]   Prasad PVV, Pisipati SR, Mutava RN, Tuinstra MR. Sensitivity of grain sorghum to high temperature stress during reproductive development. Crop Sci 2008; 48(5): 1911-7.
[http://dx.doi.org/10.2135/cropsci2008.01.0036]

[88]   Chao LM, Liu YQ, Chen DY, Xue XY, Mao YB, Chen XY. Arabidopsis transcription factors SPL1 and SPL12 confer plant thermotolerance at reproductive stage. Mol Plant 2017; 10(5): 735-48.
[http://dx.doi.org/10.1016/j.molp.2017.03.010] [PMID: 28400323]

[89]   Nadeem M, Li J, Wang M, *et al.* Unraveling field crops sensitivity to heat stress: Mechanisms, approaches, and future prospects. Agronomy (Basel) 2018; 8(7): 128.
[http://dx.doi.org/10.3390/agronomy8070128]

[90]   Matiu M, Ankerst DP, Menzel A. Interactions between temperature and drought in global and regional crop yield variability during 1961-2014. PLoS One 2017; 12(5): e0178339.
[http://dx.doi.org/10.1371/journal.pone.0178339] [PMID: 28552938]

[91]   Jumrani K, Bhatia VS. Impact of combined stress of high temperature and water deficit on growth and seed yield of soybean. Physiol Mol Biol Plants 2018; 24(1): 37-50.
[http://dx.doi.org/10.1007/s12298-017-0480-5] [PMID: 29398837]

[92]   Soba D, Arrese-Igor C, Aranjuelo I. Additive effects of heatwave and water stresses on soybean seed yield is caused by impaired carbon assimilation at pod formation but not at flowering. Plant Sci 2022; 321: 111320.
[http://dx.doi.org/10.1016/j.plantsci.2022.111320] [PMID: 35696920]

[93]   Zampieri M, Ceglar A, Dentener F, Toreti A. Wheat yield loss attributable to heat waves, drought and water excess at the global, national and subnational scales. Environ Res Lett 2017; 12(6): 064008.
[http://dx.doi.org/10.1088/1748-9326/aa723b]

[94]   Nankishore A, Farrell AD. The response of contrasting tomato genotypes to combined heat and drought stress. J Plant Physiol 2016; 202: 75-82.
[http://dx.doi.org/10.1016/j.jplph.2016.07.006] [PMID: 27467552]

[95]   Hamidou F, Halilou O, Vadez V. Assessment of groundnut under combined heat and drought stress. J Agron Crop Sci 2013; 199(1): 1-11.
[http://dx.doi.org/10.1111/j.1439-037X.2012.00518.x]

[96]   Sehgal A, Sita K, Kumar J, *et al.* Effects of drought, heat and their interaction on the growth, yield and photosynthetic function of lentil (*Lens culinaris* Medikus) genotypes varying in heat and drought sensitivity. Front Plant Sci 2017; 8: 1776.
[http://dx.doi.org/10.3389/fpls.2017.01776] [PMID: 29089954]

[97]   Ergo VV, Lascano R, Vega CRC, Parola R, Carrera CS. Heat and water stressed field-grown soybean: A multivariate study on the relationship between physiological-biochemical traits and yield. Environ Exp Bot 2018; 148: 1-11.
[http://dx.doi.org/10.1016/j.envexpbot.2017.12.023]

[98]   Fábián A, Sáfrán E, Szabó-Eitel G, Barnabás B, Jäger K. Stigma functionality and fertility are reduced by heat and drought co-stress in wheat. Front Plant Sci 2019; 10: 244.
[http://dx.doi.org/10.3389/fpls.2019.00244] [PMID: 30899270]

[99]   Zahra N, Wahid A, Hafeez MB, Ullah A, Siddique KHM, Farooq M. Grain development in wheat under combined heat and drought stress: Plant responses and management. Environ Exp Bot 2021; 188: 104517.
[http://dx.doi.org/10.1016/j.envexpbot.2021.104517]

[100] Ihsan MZ, El-Nakhlawy FS, Ismail SM, Fahad S, daur I. Wheat phenological development and growth studies as affected by drought and late season high temperature stress under arid environment. Front Plant Sci 2016; 7: 795.
[http://dx.doi.org/10.3389/fpls.2016.00795] [PMID: 27375650]

[101] Schmidt J, Claussen J, Wörlein N, *et al.* Drought and heat stress tolerance screening in wheat using computed tomography. Plant Methods 2020; 16(1): 15.
[http://dx.doi.org/10.1186/s13007-020-00565-w] [PMID: 32082405]

[102] Zhang B, Li W, Chang X, Li R, Jing R. Effects of favorable alleles for water-soluble carbohydrates at grain filling on grain weight under drought and heat stresses in wheat. PLoS One 2014; 9(7): e102917.
[http://dx.doi.org/10.1371/journal.pone.0102917] [PMID: 25036550]

[103] Pandey P, Ramegowda V, Senthil-Kumar M. Shared and unique responses of plants to multiple individual stresses and stress combinations: physiological and molecular mechanisms. Front Plant Sci 2015; 6: 723.
[http://dx.doi.org/10.3389/fpls.2015.00723] [PMID: 26442037]

[104] Koussevitzky S, Suzuki N, Huntington S, *et al.* Ascorbate peroxidase 1 plays a key role in the response of Arabidopsis thaliana to stress combination. J Biol Chem 2008; 283(49): 34197-203.
[http://dx.doi.org/10.1074/jbc.M806337200] [PMID: 18852264]

[105] Johnson SM, Lim FL, Finkler A, Fromm H, Slabas AR, Knight MR. Transcriptomic analysis of *Sorghum bicolor* responding to combined heat and drought stress. BMC Genomics 2014; 15(1): 456.
[http://dx.doi.org/10.1186/1471-2164-15-456] [PMID: 24916767]

SUBJECT INDEX